Discrete Dynamical Systems and Chaotic Machines

Theory and Applications

Jacques M. Bahi

Christophe Guyeux

CRC Press
Taylor & Francis Group
Boca Raton London New York

CRC Press is an imprint of the
Taylor & Francis Group, an **informa** business

A CHAPMAN & HALL BOOK

CHAPMAN & HALL/CRC
Numerical Analysis and Scientific Computing

Aims and scope:

Scientific computing and numerical analysis provide invaluable tools for the sciences and engineering. This series aims to capture new developments and summarize state-of-the-art methods over the whole spectrum of these fields. It will include a broad range of textbooks, monographs, and handbooks. Volumes in theory, including discretisation techniques, numerical algorithms, multiscale techniques, parallel and distributed algorithms, as well as applications of these methods in multi-disciplinary fields, are welcome. The inclusion of concrete real-world examples is highly encouraged. This series is meant to appeal to students and researchers in mathematics, engineering, and computational science.

Editors

Proposals for the series should be submitted to one of the series editors above or directly to:

CRC Press, Taylor & Francis Group
4th, Floor, Albert House
1-4 Singer Street
London EC2A 4BQ
UK

Published Titles

Classical and Modern Numerical Analysis: Theory, Methods and Practice
Azmy S. Ackleh, Edward James Allen, Ralph Baker Kearfott, and Padmanabhan Seshaiyer

Cloud Computing: Data-Intensive Computing and Scheduling
Frédéric Magoulès, Jie Pan, and Fei Teng

Computational Fluid Dynamics
Frédéric Magoulès

A Concise Introduction to Image Processing using C++
Meiqing Wang and Choi-Hong Lai

Decomposition Methods for Differential Equations: Theory and Applications
Juergen Geiser

Desktop Grid Computing
Christophe Cérin and Gilles Fedak

Discrete Dynamical Systems and Chaotic Machines: Theory and Applications
Jacques M. Bahl and Christophe Guyeux

Discrete Variational Derivative Method: A Structure-Preserving Numerical Method for Partial Differential Equations
Daisuke Furihata and Takayasu Matsuo

Grid Resource Management: Toward Virtual and Services Compliant Grid Computing
Frédéric Magoulès, Thi-Mai-Huong Nguyen, and Lei Yu

Fundamentals of Grid Computing: Theory, Algorithms and Technologies
Frédéric Magoulès

Handbook of Sinc Numerical Methods
Frank Stenger

Introduction to Grid Computing
Frédéric Magoulès, Jie Pan, Kiat-An Tan, and Abhinit Kumar

Iterative Splitting Methods for Differential Equations
Juergen Geiser

Mathematical Objects in C++: Computational Tools in a Unified Object-Oriented Approach
Yair Shapira

Numerical Linear Approximation in C
Nabih N. Abdelmalek and William A. Malek

Numerical Techniques for Direct and Large-Eddy Simulations
Xi Jiang and Choi-Hong Lai

Parallel Algorithms
Henri Casanova, Arnaud Legrand, and Yves Robert

Parallel Iterative Algorithms: From Sequential to Grid Computing
Jacques M. Bahi, Sylvain Contassot-Vivier, and Raphael Couturier

Particle Swarm Optimisation: Classical and Quantum Perspectives
Jun Sun, Choi-Hong Lai, and Xiao-Jun Wu

XML in Scientific Computing
C. Pozrikidis

CRC Press
Taylor & Francis Group
6000 Broken Sound Parkway NW, Suite 300
Boca Raton, FL 33487-2742

First issued in paperback 2019

© 2013 by Taylor & Francis Group, LLC
CRC Press is an imprint of Taylor & Francis Group, an Informa business

No claim to original U.S. Government works

ISBN-13: 978-1-4665-5450-4 (hbk)
ISBN-13: 978-0-367-37994-0 (pbk)

Visit the Taylor & Francis Web site at
http://www.taylorandfrancis.com

and the CRC Press Web site at
http://www.crcpress.com

Contents

List of Figures

List of Tables

Preface

The purpose of this book is to introduce so-called discrete dynamical systems and to present the latest developments in this field. Their chaotic behaviors and their use in computer science are investigated. This book details the latest results obtained by the community working in this domain, investigating not only the convergence of such systems, but the operation of divergence, random, or chaotic behaviors too.

Such behaviors have already been well-studied within the framework of the mathematical theory of chaos, which are recalled in the beginning of this book. However, until the recent researches of the authors in this field, the practical implementation (on finite machines) of this theory raised several problems.

During these past years, the authors have proposed an original and innovative approach, proving that it is absolutely possible to make finite machines (computers, neural networks, wireless sensor networks...) work chaotically, in the most rigorous understanding of this term. The key idea is to take into account the fact that these machines do not work in a vacuum, but have to interact with the outside world.

The authors have used these deterministic machines with unpredictable behaviors in the fields of security, modeling, and numerical simulations. These original works have been validated by the scientific community, leading to almost forty publications in peer-reviewed international conferences and journals.

These works are presented here rigorously and in greater detail, concerning both theoretical and practical aspects. This book, which is self-contained, presents the following elements:

- the key notions in chaos theory, under their mathematical and physical formulations;

- fundamentals in computer science, establishing clearly that these chaos properties can be satisfied by finite state machines;

- concrete realization of chaotic machines in the computer science security field (pseudorandom number generators, hash functions, digital watermarking, and steganography) and in wireless sensor networks.

In the computer science security field, this theory allows the study of security problems that could not be tackled until now (as in steganalysis). It can also reinforce the confidence put in existing cryptographically secure

xvi

tools. We show that it is possible to preserve cryptographic properties while adding some relevant chaotic behaviors, and we explain why such properties are indeed interesting ones. Concrete illustrations are proposed in the fields of pseudorandom number generators (PRNGs), hash functions, and digital watermarking.

A similar approach is used for the applications to wireless sensor networks (WSN). Two fundamental concerns will be addressed, namely secure data aggregation and video-surveillance. Explanations are given on how to use chaos in order to enlarge the number of contexts within which security can be evaluated, and how to reinforce the security of existing schemes.

Due to its intrinsically multidisciplinary nature, this book may be of interest to post-secondary students, engineers, or researchers in the fields of mathematics, computer science, physics, or biology, who have an interest in complex and chaotic phenomena and their use, study, modeling, and computing.

Symbol Description

Symbol Description

$\mathbb{B}$ The Boolean set $\{0, 1\}$.

$\mathbb{N}$ The natural numbers.

$[\![1; N]\!]$ The interval of integers $\{1, 2, \ldots, N\}$.

$\mathbb{R}$ The real numbers.

$\mathcal{P}(X)$ The set of subsets of X.

$X \times Y$ The Cartesian product of sets X, Y.

S^n The nth term of a sequence S.

V_i The ith component of a vector V.

f^k $= f \circ \ldots \circ f$ denotes the kth composition of a function f.

$\liminf\limits_{n \to \infty} x_n$ The limit inferior of (x_n).

$\limsup\limits_{n \to \infty} x_n$ The limit superior of (x_n).

Part I

An introduction to chaos

Chapter 1

Classical Examples by Way of Introduction

Let us start this book by a short initiation to chaos through two archetypes inspired by the work of Ian Stewart [129].

1.1 Historical Context

Recurrent sequences, also called discrete dynamical systems, of the form

$$u^0 \in \mathbb{R}, u^{n+1} = f(u^n), \tag{1.1}$$

with f continuous, have been well studied since the early years of mathematical analysis. They are widely used to resolve equations using a Newtonian method, or when approximating the solutions to differential equations using finite difference equations to approximate derivatives. The context study was the seek for convergence, which is for instance guarantee when using monotonic functions or contractions. In the middle of the last century, Coppel established a link between this desire of convergence and the existence of a cycle in iterations [57]. More precisely, his theorem states that, considering Eq. (1.1) with

3

a function $f : I \longrightarrow I$ continuous on the line segment I, the absence of any 2-cycle implies the convergence of the discrete dynamical system.

    This theorem establish a clear link between the existence of a cycle of a given length and the convergence of the system. In other words, between cycles and order. Conversely, Li and Yorke established in 1975 [92] that the presence of a point of period three implies chaos in the same situation as previously. By chaos, they mean the existence of points of any period: this kind of disorder, which is the first occurrence of the term "chaos" in the mathematical literature, is thus related to the multiplicity of periods. Since that time, the mathematical theory of chaos has known several developments to qualify or quantify the richness of chaos presented by a given discrete dynamical system, one of the most famous works, although old, being the one by Feigenbaum.

1.2   Feigenbaum's Bifurcation

1.2.1   An iterative process

    Consider a real number between 0 and 1, on which the function $x \mapsto x^2$ is iterated. In other words, let us compute the terms of the sequence

$$\begin{cases} x_0 \in [0, 1] \\ x_{n+1} = x_n^2. \end{cases} \qquad (1.2)$$

    A regular behavior can be observed: a convergence to 0 (see Fig. 1.1). Realizing a similar operation replacing the iteration function $f(x) = x^2$ by $f(x) = cos(x)$ leads another time to an ordered behavior, namely a convergence to 0.739085133 (Fig. 1.2).

    When considering the *tan* function, 0.54321, as the initial value, and a precision of 8 digits, another convergence seems to appear. However, the same iteration with improved precision leads to a behavior similar to a slow divergence to $+\infty$.

    A first statement is that the way to observe a phenomenon is of importance: without a library of large numbers, the convergence of iterations was the most probable behavior, whereas a divergence to $+\infty$ appears when using such a library. However, with or without the library, the behavior of the iterations still remains ordered, and predictable if it is well observed: a slow growth to $+\infty$.

    Iterating other functions as the exponential or the square root leads to a similar convergence, whereas iterations of the inverse function implies a periodic behavior of period 2. Trying several other reference functions leads to one of these three ordered behaviors: convergence, divergence to $\pm\infty$, or periodicity.

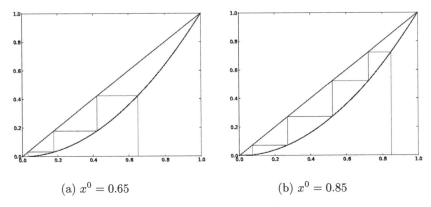

(a) $x^0 = 0.65$ (b) $x^0 = 0.85$

FIGURE 1.1: Iterations of $f(x) = x^2$

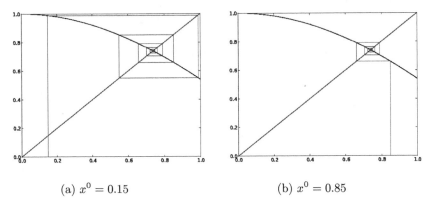

(a) $x^0 = 0.15$ (b) $x^0 = 0.85$

FIGURE 1.2: Iterations of $f(x) = cos(x)$

A first idea is that this reasonable behavior comes from the fact that only reference functions have been iterated. However the function $x^2 - 1$ leads another time to a periodic behavior (of period 2). However the function $2x^2 - 1$, with an initial value between 0 and 1, breaks this illusion of regularity.

The function $2x^2 - 1$ is simple and similar to other functions that have been used below. However, no regularity appears when iterating it. Results appear random, and the order met in previous examples is not there.

Let us compare the evolution of the system for two very close initial values: 0.54321 and 0.54322. After fifty iterations, completely different behaviors can be observed, even if points was originally close. The same work with more close initial values and improved precision leads to the same conclusion, that is, a strong sensitivity to the initial conditions.

Let us study the behaviors of the function $x \mapsto kx^2 - 1$, with k between

1 and 2. As k increases, cycles of increasing lengths appear (period 16 for $k = 1.4$), and a chaotic behavior appears for $k = 1.5$. Then, increasing k implies a more chaotic behavior.

Consider for instance the behavior of the system for $k = 1.74$ and $k = 1.75$. For 1.74, a well developed chaos is initiated, whereas for $k = 1.75$, the sequence enters into a cycle of length 3. Order and chaos are then intrinsically linked, and order appears from chaos. Let us study more systematically these iterations.

1.2.2   Feigenbaum's bifurcation

1.2.2.1   Presentation

The model of the previous section, particularly simple, possesses almost all the complex properties that can be found in dynamical systems. This model is called *Feigenbaum's bifurcation*. It consists more exactly in the transformation of the interval $[-1; 1]$, which transforms the point x into the point $1 - \mu x^2$, where μ is a parameter chosen between 0 and 2.

1.2.2.2   Parameter μ between 0 and 0.75

Let us first study $\mu \in [0; 0.75]$: for $x_0 = 0$ and for $x_0 = -0.5$, there is a convergence to 0.732050807. Indeed, whatever the initial value x_0, the natural evolution of the system conducts it to the point 0.732050807. This point 0.732050807 is thus a stable equilibrium of the system for $\mu = 0.5$.

It is possible to prove that,

Theorem 1 *For all μ between 0 and 0,75, Feigenbaum's bifurcation possesses a stable unique equilibrium $\bar{x}$, whose exact position is continuously dependent on μ.*

More precisely, $\bar{x}$ is the solution of the system $\mu x^2 + x - 1 = 0$.

1.2.2.3   Parameter μ between 0.75 and 1.25

Consider $\mu \in [0, 75; 1, 25]$. For any initial value x_0, a 2-periodic trajectory appears, that alternatively goes from 0 to 1. The point $\bar{x}$ obtained previously still remains an equilibrium point: if $x_0 = \bar{x}$, then $x_n = \bar{x}, \forall n$. This equilibrium is however unstable: if the initial value is closed but different from $\bar{x}$, then the 2-cycle reappears another time.

1.2.2.4   Parameter between 1.25 and 1.368

Consider $\mu = 1, 3$. A trajectory of period 4 can be observed, and all other trajectories tend to this latter. $\bar{x}$ is again an unstable equilibrium, and there is also a 2-periodic trajectory, unstable, going by the points

$$\frac{1}{2\mu} \left(1 + \sqrt{4\mu - 3} \right) \text{ and } \frac{1}{2\mu} \left(1 - \sqrt{4\mu - 3} \right) \tag{1.3}$$

1.2.2.5 Summary

Two "catastrophes" have been discovered:

- When μ belongs in $[0; 0.75]$, the qualitative behavior of the system does not change: it presents a stable equilibrium, moving continuously with μ, and all the trajectories are convergent to it.

- When μ belongs in $[0, 75; 1, 25]$, the behavior does not change: a 2-periodic stable trajectory appears, and all other trajectories are convergent to it.

However, the threshold $\mu = 0.75$ changes the qualitative behavior: the stable equilibrium is removed to the benefit of a 2-periodic trajectory. Similarly, $\mu = 1.25$ is a catastrophic value, as the 2-periodic trajectory loses its stability to the profit of a 4-periodic trajectory.

1.2.2.6 Parameter between 1.368 and 1.401

For μ between 1.368 and 1.401, successive doubling periods appear. More precisely, there is an infinite sequence of "catastrophic" thresholds μ_n:

- that increase to 1.401,

- such that if $\mu \in [\mu_n; \mu_{n+1}]$, then the system has a stable trajectory of period 2^{n+1}, which is such that any other trajectory is convergent to it.

Thus going through these catastrophic thresholds in the sense of increasing μ corresponds to a period doubling. It is possible to state with a good approximation that:

$$1.401 - \mu_n = \text{Constant} \times 4.6692...^{-n} \qquad (1.4)$$

or

$$\frac{1.401 - \mu_n}{1.401 - \mu_{n+1}} = 4.6692... \qquad (1.5)$$

Leading to the definition:

Definition 1 (Feigenbaum Constant) Number 4.6692... is called the *Feigenbaum constant*.

This constant, which is now known with good accuracy, appears in a lot of other circumstances: it seems to present a profound signification, that can be related to cascade bifurcation phenomena.

1.2.2.7 Parameter between 1.401 and 2

The area $\mu \in [1.401; 2]$ still remains unknown. For most of the μ values, the system presents a chaotic behavior: all the trajectories that can be found are unstable, and the system evolves in a random manner, crossing the whole interval.

However, in this reign of disorder, small areas of stability and order occur, like the interval $1.75 < \mu < 1.7685$.

1.3   The Logistic Map

1.3.1   Definition of the logistic map

Let us now introduce another function whose iterates behave as the Feigenbaum bifurcation. This is the so called "logistic maps", a famous example of chaos.

Definition 2 The logistic map is defined by $\forall x \in [0,1], f(x) = ax(1-x)$, where a is a parameter chosen into $[0,4]$.

Another time, we focus on the dynamical system $u_0 \in [0;1], u_{n+1} = f(u_n)$, which can be represented graphically, as in Figure 1.3.

1.3.2   Behavior of the sequence

The dynamical system of the logistic map has a variable behavior, depending on the value of parameter a. As it is depicted in Figure 1.3,

- For $a < 3$, the sequence is convergent.

- For a slightly greater than 3, the stable point of f loses its attractive property. A 2-cycle appears (two stable points for $f \circ f$ in addition to the stable point of f), that is, a periodic orbit.

- Orbits of period 4, 8, etc. appear when increasing the value of the parameter a.

For $a \simeq 3,86$, as shown in Figure 1.4, the orbit of the sequence presents an erratic character, completely different from the previous graphics that where predictable.

1.3.3   Sensitivity to the initial conditions

For two values different from 0.1%, by obtained after 25 iterations differ from 80%, as it is depicted in Figure 1.5.

Thus the logistic map, which is fully known and easily definable (fully deterministic) has an unpredictable behavior. More exactly, a slight variation of the initial value can possibly change a convergence into a periodic one, or a periodic behavior into a random one, which is sometimes referred as the "butterfly effect."

This butterfly effect comes from meteorology, meaning that the tomorrow's weather depends on today's climatic conditions. However, it is impossible to measure these current conditions with sufficient precision. As meteorology is ruled by a chaotic system, a slight error in current data can possibly lead to completely different predictions. This point is detailed in the next section.

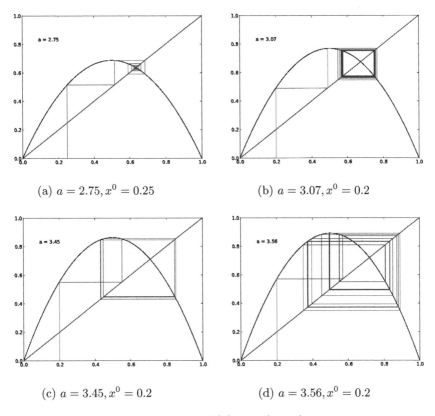

(a) $a = 2.75, x^0 = 0.25$

(b) $a = 3.07, x^0 = 0.2$

(c) $a = 3.45, x^0 = 0.2$

(d) $a = 3.56, x^0 = 0.2$

FIGURE 1.3: Iterations of $f(x) = ax(1 - x), a < 3.86$

1.4 The Lorenz System

Let us introduce another classical system in the mathematical theory of chaos.

In 1963, the meteorologist Edward Lorenz for the first time stated the chaotic character of meteorology. Mathematically speaking, interactions between atmosphere and ocean are described by the Navier-Stokes equation of fluid mechanics. This system of equations was too complicated to solve numerically.

The latter thus studies a very simplified model, to investigate a particular physical situation (the convection phenomenon of Rayleigh-Bénard). Such a study leads to a differential dynamical system having only three degrees of freedom, easier to integrate the former equations.

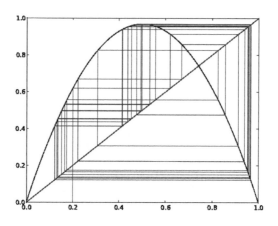

FIGURE 1.4: The logistic map, with $a = 3.87$

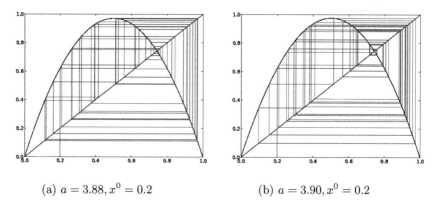

(a) $a = 3.88, x^0 = 0.2$ (b) $a = 3.90, x^0 = 0.2$

FIGURE 1.5: Iterations of $f(x) = ax(1 - x), a \in \{3.88, 3.90\}$

This differential system is

$$\begin{cases} \frac{dx(t)}{dt} = \sigma\big(y(t) - x(t)\big) \\ \frac{dy(t)}{dt} = \rho\,x(t) - y(t) - x(t)\,z(t) \\ \frac{dz(t)}{dt} = x(t)\,y(t) - \beta\,z(t) \end{cases}$$

To obtain a graphical representation, this system must be discretized as follows:

$$\begin{cases} x_{i+1} & = & \delta t\sigma(y_i - x_i) + x_i \\ y_{i+1} & = & \delta t(rx_i - y_i - x_i z_i) + y_i \\ z_{i+1} & = & \delta t(-bz_i + x_i y_i) + z_i \end{cases}$$

FIGURE 1.6: The strange attractor of Lorenz

Taking $\sigma = 10$, $r = 28$, $b = 8/3$; $\delta t = 0.01$, and $(x_0, y_0, z_0) = (8, 3, 4)$ leads to a strange attractor having the form of a butterfly, represented below.

For almost all the initial conditions (different from the fixed points), the orbit of the system follows a trajectory on a wing, then goes on to another one, and so on, in an erratic manner.

In 1972, Edward Lorenz presented the so-called butterfly effect in the American Association for the Scientific Progress. His question still remains famous: "Does the flap of a butterfly's wings in Brazil set off a tornado in Texas? " The butterfly, instead of the ladybug, was chosen to present a parallel with the form of the strange attractor.

Lorenz claimed that it was not possible to predict correctly the climatic modifications in long term. Indeed, an uncertitude of 1 on 10^6 during the storage of the initial situation can lead to a totally erroneous prediction: two points extremely close to each other, on the attractor, could have totally different behaviors. However,

- these uncertainties cannot be avoided,

- all the elements that constitute our environment cannot be taken into account, especially small variations.

Lorenz has thus discovered that a very complex dynamic can appear in a system which is formally simple. The understanding of what is simple and what is complex has thus been fundamentally rethought. In particular, complexity can be intrinsically related to a given system, and not to external, accidental causes. Infinitesimal variations between two initial conditions can lead to totally different final solutions.

Part II

The mathematical theory
of chaos

Chapter 2

Definitions and Notations

Some definitions of mathematical topology are recalled in this chapter that will be used in a further chapter to define chaotic systems. At the same time we will begin to address the topological approach of discrete dynamical systems. Readers that desire to deepen the mathematical topology are referred to the book of Laurent Schwartz [124].

2.1 Topological and Metrical Spaces

In this section, categories of spaces of interest are recalled, as well as functions used in these spaces.

2.1.1 Topological spaces, open sets, and neighborhoods

2.1.1.1 First definitions

Let us first introduce basic topological terminologies.

Definition 3 (Topological space) A *topological space* is a couple (E, τ), where E is a set and τ a family of subsets of E, called *open sets*, verifying:

- $\varnothing, E \in \tau$: the empty set and E are open sets,

- any union of open sets is open,

- any finite intersection of open sets is open.

Definition 4 (Closed set) A *closed set* is the complement of an open set.

Let A be a subset of a topological space (E, τ). The smallest closed set containing A always exists: this is the intersection of all the closed sets that contain A. We denote it $\overline{A}$, and we speak about the *closure* of A.

Various topologies can be attached to a given set, which leads to the following order notion.

Definition 5 (Comparison of topologies) A topology $\tau \in \mathcal{P}(\mathcal{X})$ on the set $\mathcal{X}$ is *finer* than another topology $\tau' \in \mathcal{P}(\mathcal{X})$ when $\tau' \subset \tau$. τ' is said to be *coarser*, or *weaker*, than τ.

A topology is finer than another one if it contains more open sets. This is a partial order: numerous couples of topologies cannot be compared together. Let us now introduce the notion of neighborhood.

Definition 6 (Neighborhood) Let (E, τ) a topological space. We call the *neighborhood of $x \in E$* any subset of E containing an open set that contains x.

2.1.1.2   Examples of topologies

Let us give two basic examples of topologies on a given set $\mathcal{X}$.

Definition 7 (Discrete topology) The *discrete topology* of a set $\mathcal{X}$ is the topology $\tau = \mathcal{P}(\mathcal{X})$ of all the subsets of $\mathcal{X}$.

The discrete topology is the topology on $\mathcal{X}$ having the largest number of open sets. It is finer than any other topology on $\mathcal{X}$ and is such that any subset of $\mathcal{X}$ is both open and closed. This topology gets its name from the fact that all the points are isolated (each point is both open and closed). On the contrary,

Definition 8 (Trivial topology) The *trivial topology* on a set $\mathcal{X}$ is the topology $\tau = \{\varnothing, \mathcal{X}\}$.

This is the coarsest topology that can be defined on $\mathcal{X}$, having only two open sets.

2.1.2   Distances, metrical spaces

A convenient way to define topologies is to use metrics.

Definition 9 (Distance) Given a set E, a *distance*, also called a *metric*, is an application $d : E \times E \to \mathbb{R}^+$ having the following properties:

Symmetry: $\forall x, y \in E, d(x, y) = d(y, x)$.

Identity of indiscernibles: $\forall x, y \in E, d(x,y) = 0 \Leftrightarrow x = y$.

Triangle inequality: $\forall x, y, z \in E, d(x,z) \leqslant d(x,y) + d(y,z)$.

Definition 10 (Metric spaces) A *metric space* is any couple (E,d), where E is a set and d a distance on E.

Metric spaces are topological spaces. They possess particular neighborhoods called balls:

Definition 11 (Open ball, closed ball) Let (E,d) be a metric space. The *closed ball* of radius $r > 0$ and centered at a point P, is the set $\overline{\mathcal{B}}(P,r)$ of points having a distance to P lower than or equal to r: $\overline{\mathcal{B}}(P,r) = \left\{ M \in E \ / \ d(M,P) \leqslant r \right\}$.

The open ball $\mathcal{B}(P,r)$ is the set $\left\{ M \in E \ / \ d(M,P) < r \right\}$.

2.2 Compactness and Completeness

2.2.1 Compactness

Particular topological spaces called *compact spaces* will play an important role in the remainder of this book. They are defined in what follows.

2.2.1.1 Preliminaries

Notions of separated space and of open covers must be introduced first, to be able to define the compactness property.

Definition 12 (Separated Spaces) A *separated space* is a topological space in which two distinct points always possess two disjoint neighborhoods.

For instance, any metric space is separated.

Definition 13 (Cover, open cover) A *cover* of a set $\mathcal{X}$ is a set $\mathcal{P}$ of nonempty subsets of $\mathcal{X}$ such that the union of these subsets is $\mathcal{X}$. An *open cover* is a cover such that all the subsets are open.

2.2.1.2 Compactness of topological and metric spaces

Definition 14 (Compact spaces) A separated space is *compact* if each of its open covers has a finite subcover.

There is a sequential, more practicable characterization of compactness for metric spaces.

Proposition 1 (Sequential characterization) *A metric space $(\mathcal{X},d)$ is compact if and only if every sequence in the space has a convergent subsequence.*

2.2.2 Completeness

Complete spaces are metric spaces such that particular sequences, called Cauchy sequences, are convergent.

Definition 15 (Cauchy Sequence) A sequence $(x^n)_{n\in\mathbb{N}}$ of a metric space (E, d) is said to be a *Cauchy sequence* if for all real number $\varepsilon > 0$, there exists a positive integer N such that for all integers $p, q \geqslant N$, the distance $d(x^p, x^q)$ is lower than ε : $\forall \varepsilon > 0, \exists N \in \mathbb{N}, \forall p, q > N, d(x^p, x^q) < \varepsilon$.

This is a sequence whose elements become arbitrarily close to each other as the sequence progresses.

Definition 16 (Complete metric space) A metric space (E, d) is said to be *complete* if any Cauchy sequence of (E, d) has a limit into (E, d).

Intuitively, a space is complete if there are no "points missing" from it (inside or at the boundary). For instance, the set of rational numbers is not complete, because e.g. $\sqrt{2}$ is "missing" from it, even though one can construct a Cauchy sequence of rational numbers that converges to it.

2.3 Continuity

The continuity notion can be defined in topological spaces in the following manner.

Definition 17 (Continuity) Let f be an application between two topological spaces. It is *continuous in x* if for any neighborhood V of $f(x)$, there exists a neighborhood V' of x whose image $f(V')$ is in V.

Continuity can be formulated more easily when considering applications between metrical spaces.

Definition 18 (Continuity) Let (E, d) and (E', d') be two metrical spaces, $a \in E$ and $f : E \to E'$. The application f is *continuous in a* if:

$$\forall \varepsilon > 0, \exists \eta > 0, \forall x \in E, d(x, a) \leq \eta \implies d'(f(x), f(a)) \leq \varepsilon.$$

In metrical spaces, continuity can be proven easily by using the following sequential characterization.

Proposition 2 *Let $f : (E, d) \to (E', d')$ be an application between two metric spaces.*
Then f is continuous in $a \in E$ if and only if, for any sequence x^n which converges to a, the sequence $f(x^n)$ converges to $f(a)$.

2.4 Discrete Dynamical Systems

Let $f : \mathcal{X} \longrightarrow \mathcal{X}$ be an application from a topological or metrical space $\mathcal{X}$ to $\mathcal{X}$.

Consider the sequence defined by the recurrent relation:

$$\begin{cases} x^0 \in \mathcal{X} \\ \forall n \in \mathbb{N}, x^{n+1} = f(x^n). \end{cases}$$

The behavior of these iterations depends on the function f and on the "phase space" $\mathcal{X}$ on which we iterate, leading to the definition [66] :

Definition 19 (Discrete dynamical system) A *discrete dynamical system* is a couple $(\mathcal{X}, f)$ constituted by:

- a nonempty topological space $(\mathcal{X}, \tau)$ called *phases space*,

- a continuous function $f : \mathcal{X} \longrightarrow \mathcal{X}$ called *iteration function*.

Function f can be bijective, which allows to invert the system. If so, this system is said to be *reversible* [66]:

Definition 20 (Reversible discrete dynamical system) A discrete dynamical system $(\mathcal{X}, f)$ is *reversible* if f is a (topological) homeomorphism, i.e., if f is a bicontinuous bijection.

An important question is to determine the way a given point x evolves as the iterations pass, that is the form of the orbit of x:

Definition 21 (Orbit) For a given $x \in \mathcal{X}$, the sequence $(f^{(n)}(x))_{n \in \mathbb{N}}$ is called *orbit* of x. It is denoted by $\gamma_x = (\gamma_x^n)_{n \in \mathbb{N}}$, where $\gamma_x^n = f^n(x)$.

These orbits will be studied more deeply in the next section.

Chapter 3

Devaney's Formulation of Chaos

The mathematical theory of chaos has known a lot of developments since the first occurrence of this term in 1975 [92], one of the most known being the definition of chaos given by Robert L. Devaney [61]. The goal of this chapter is to introduce his definition, whereas other definitions will be given in the next chapter. The reader who wants to go deeper into the topic is referred to

the book by Devaney [61], and to the theses of Enrico Formenti [66], [67] and
Sylvie Ruette [123].

3.1   Periodicity, Stability, and Regularity

The theory of chaos tries to determine whether the behavior of a discrete
dynamical system can be predicted or not. That is to say, if it is possible to
guess what will be the orbit γ_x of a given point x. In that sense, points whose
behavior is the easiest to grasp are periodic and stable points.

3.1.1   Periodic and stable points

Definition 22 (Periodic point) A point $p \in \mathcal{X}$ is said to be *periodic*
of *period* k if $k \neq 0$ is a natural number such that $f^{(k)}(p) = p$ and
$\forall h \in [\![0; k-1]\!], f^{(h)}(p) \neq p$.

$Per_k(f)$ denotes the set of k-periodic points of f, and $Per(f)$ the set of all
periodic points (whatever their period). Periodicity can occur after a more or
less large transitional phase, which leads to a variant of the previous definition.

Definition 23 (Ultimately periodic point) A point is said to be *ultimately periodic* if there exist two natural numbers n and p such that
$f^{(n+p)}(x) = f^{(p)}(x)$.
Let n_0 be the smallest n verifying this property. The set $\{x, f(x), \dots, f^{(n_0)}(x)\}$
is called *transient* of x and n_0 is the *length of the transient*.

Stable points are points having the simplest orbit.

Definition 24 (Stable point) Periodic points of period 1 are called *fixed
points* of f, or *stable points*. Similarly, ultimately periodic points of period 1
are called *ultimately fixed points*.

3.1.2   Regular systems

In mathematical topology, the density concept for a subset A of a topolog-
ical space $\mathcal{X}$ allows to reflect the idea that for all point x of $\mathcal{X}$, it is possible
to find another point of A being as close as possible to x.

Definition 25 (Dense space) Let $\mathcal{X}$ be a topological space and A a subset
of $\mathcal{X}$. A is said to be *dense* in $\mathcal{X}$ if for any element x of $\mathcal{X}$, any neighborhood
of x contains at least one point of A.

It is now possible to define a first aspect of chaos, considering periodic
points [66]:

Definition 26 (Regular systems) A discrete dynamical system $(\mathcal{X}, f)$ is said to be *regular* if the set of periodic points for f is dense in $\mathcal{X}$.

Remark 1 In a metrical space $(\mathcal{X}, d)$, the dynamical system $(\mathcal{X}, f)$ is regular if and only if $\forall x \in \mathcal{X}, \forall \varepsilon > 0, \exists p \in Per(f), d(x, p) < \varepsilon$.

Even if the term "regular" seems to be in contradiction with the idea of chaos, we should have in mind that this definition can lead to a certain type of unpredictability. Indeed, if numerical simulations of the evolution of the system are realized, small errors in initial conditions *can* lead to an orbit radically different from the desired one. For instance, it is possible to obtain a large period or a non-periodic orbit, instead of an orbit having a small period, as stated in Chapter 1.

This regularity cannot define alone a notion of chaos, as pathological situations can possibly occur. Furthermore, elementary systems are chaotic under that conception, like the identity map, permutations, and so on. For these reasons, elements of regularity are generally coupled with a property of non decomposability.

3.2   Simplification of Discrete Dynamical Systems

A central point in the study of dynamical systems is to characterize their asymptotic behaviors. To focus on such behaviors leads one to study the subsets of the phase space that are stable under the action of the system.

3.2.1   Invariant subsystems

Definition 27 (Positive non-variance) Let $(\mathcal{X}, f)$ be a discrete dynamical system. A subset A of $\mathcal{X}$ is said to be *positively invariant* if $f(A) \subset A$, and *strictly positively invariant* in case of equality.

Example 1 The subsets $Per(f)$, $Per_k(f)$ are positively invariants.

A positively invariant set is thus a "caption set": once entering into this subset, the iterations will never exit it [66]. If $A \subset \mathcal{X}$ is positively invariant, then obviously:

$$\forall n \in \mathbb{N}, f^n(A) \subset A.$$

The couple (A, f) can thus be considered as a dynamical subsystem of $(\mathcal{X}, f)$. In other words, it is possible to study separately the positively invariant subsets of $\mathcal{X}$. They are smaller than the whole system, thus they are probably easier to understand. It is possible to prove that:

Proposition 3 *If A is positively invariant, then its topological closure $\overline{A}$ is invariant too.*

3.2.2 (Un)decomposability and transitivity

Some systems are easy to study, because they can be integrally decomposed into positively invariant subsystems. This is not the case for transitive systems.

3.2.2.1 Decomposability

Decomposable systems are systems admitting some particular covers (*c.f.* Definition 13) that evolve in a predictable manner under the action of f.

Definition 28 (Decomposability) A discrete dynamical system $(\mathcal{X}, f)$ is *decomposable* if there exists a finite open cover (having at least two elements) of $\mathcal{X}$ such that each open set of the cover is a positively invariant set of f.

Independent dynamical subsystems are then acting on each element of the cover. Consequently, each open set can be studied separately, and the complete behavior of $(\mathcal{X}, f)$ can be deduced secondly. By doing so, the study of the whole system is thus simplified.

3.2.2.2 Transitivity

Transitivity is the opposite of the property of decomposability.

Definition 29 (Transitivity) A discrete dynamical system $(\mathcal{X}, f)$ is said to be *transitive* if for each couple of open nonempty sets $A, B \subset \mathcal{X}$, there exists $k \in \mathbb{N}$ such that $f^{(k)}(A) \cap B \neq \varnothing$.

Transitivity implies indecomposability, an irreductibility condition that is defined as follows [66]:

Definition 30 (Indecomposability) A discrete dynamical system $(\mathcal{X}, f)$ is *indecomposable* if and only if it is not the union of two nonempty subsets, closed, and positively invariant.

Let us now give some consequences of being transitive [66, 67], to have a better understanding of this notion, which is a fundamental one in the mathematical theory of chaos.

3.2.2.3 Consequences of the transitivity property

First of all, when a discrete dynamical system is transitive, it is possible to be as closed as desired of each $x \in \mathcal{X}$ when starting to iterate from any open set of $\mathcal{X}$. More precisely,

Proposition 4 *A discrete dynamical system $(\mathcal{X}, f)$ is transitive if and only if for any open nonempty set A,*

$$\overline{\bigcup_{n \in \mathbb{N}} f^n(A)} = \mathcal{X}$$

In other words, a system is transitive if it is possible to join any open set starting from any other open set:

Proposition 5 *A discrete dynamical system* $(\mathcal{X}, f)$ *is transitive if and only if for each couple* $(x, y) \in \mathcal{X}^2$, *and for any balls* B_x, B_y *respectively centered in* x *and* y, *we have:*

$$\exists z \in B_x, \exists n_0 \in \mathbb{N}, f^{(n_0)}(z) \in B_y.$$

Finally, transitive systems visit the whole space; they will forget no location [66]:

Proposition 6 *If* $(\mathcal{X}, f)$ *is transitive, then* $\overline{f(\mathcal{X})} = \mathcal{X}$.

PROOF Let $x \in \mathcal{X}$. If there exists a ball B_x such that $B_x \cap f(\mathcal{X}) = \varnothing$, then let U an open set that is disjoint from B_x. By transitivity, $\exists n > 0$ such that:

$$\varnothing \neq B_x \cap f^n(U) \subset B_x \cap f^n(\mathcal{X}) \subset B_x \cap f(\mathcal{X}) = \varnothing,$$

which is absurd.

3.2.2.4 Transitivity in compact spaces

We recall in this section some particular results in the case where $(\mathcal{X}, f)$ is a transitive discrete dynamical system with $(\mathcal{X}, d)$ compact. The reader will find the proofs of these results in [66, 67, 123].

First of all, the transitivity property has an equivalent formulation in compact spaces:

Proposition 7 *In compact spaces, transitivity is equivalent to possessing a dense orbit.*

In other words, as the space is relatively small (compact), transitivity means that there exists at least a point that will visit "almost" all the space. Additionally, in case of compactness, f is necessarily onto (surjective):

Proposition 8 *Let* $(\mathcal{X}, f)$ *a transitive discrete dynamical system, such that* $(\mathcal{X}, d)$ *is compact. Then* $f(\mathcal{X}) = \mathcal{X}$.

Finally, properties of Section 3.2.2.3 have the following reformulation in compact spaces.

Proposition 9 *Let* $(\mathcal{X}, f)$ *a discrete dynamical system, with* $\mathcal{X}$ *compact. The following properties are equivalents:*

1. $(\mathcal{X}, f)$ *is transitive.*

2. *For all open nonempty sets* A, B *and for all* $k \in \mathbb{N}$, $f^{-k}(A) \cap B \neq \varnothing$.

3. *For any nonempty open set* A, $\overline{\cup_{n \in \mathbb{N}} f^{-n}(A)} = \mathcal{X}$.

3.2.3 Stronger formulations of transitivity

There exist stronger versions of transitivity that lead to a larger unpredictability.

3.2.3.1 Total transitivity

Some dynamical systems can be easier to understand by studying subsequences of $x^n = f^n(x^0)$, like evaluating the odd and even terms of the sequence separately. This is why total transitivity extends the property of transitivity at all the subsequences of $x^n = f^n(x^0)$.

Definition 31 (Total transitivity) f is said to be *totally transitive* when $\forall n \geqslant 1$, the composition map f^n is transitive.

3.2.3.2 Strong transitivity

Definition 32 (Strong transitivity) A discrete dynamical system $(\mathcal{X}, f)$ is said to be *strongly transitive* if

$$\forall x, y \in \mathcal{X}, \forall r > 0, \exists z \in B(x, r), \exists n \in \mathbb{N}, f^n(z) = y.$$

In other words, for each couple x, y, there exists a point as close as desired to x that has an iterate *equal to* y: in addition to being able to reach any other open set, it is demanded here to reach any other point.

Let us remark the following result [66]:

Proposition 10 *In compact metric spaces, transitivity and strong transitivity are equivalent.*

3.2.3.3 Topological mixture

The topological mixture is another strong version of transitivity.

Definition 33 (Topological mixture) A discrete dynamical system is *topologically mixing* if and only if for all couple of disjoints nonempty open sets U and V, there exists $n_0 \in \mathbb{N}$ such that $\forall n \geqslant n_0, f^n(U) \cap V \neq \varnothing$.

3.2.4 Perfect discrete dynamical systems

Let us finally introduce the definition of some particular spaces called *perfects*, in which the transitivity property is obtained when a point with a dense orbit has been discovered.

Definition 34 (Accumulation point) A point x of a topological space (E, τ) is an *accumulation point* of a subset $A \subset E$ if any neighborhood of x contains an infinity number of points of A.

Definition 35 (Perfect systems) A discrete dynamical system $(\mathcal{X}, f)$ is said to be *perfect* if each point of $\mathcal{X}$ is an accumulation point in $\mathcal{X}$.

It is possible to prove that:

Proposition 11 *If $(\mathcal{X}, f)$ is perfect and possesses a dense orbit, then it is transitive.*

The reciprocal proposition is interesting too [66]:

Proposition 12 *Let us suppose that the metrical space $(\mathcal{X}, d)$ is compact. If $(\mathcal{X}, d)$ is transitive, then it has a dense orbit.*

PROOF If $(\mathcal{X}, d)$ is compact, then it is separated. Let $(U_1, \ldots, U_n) \subset \mathcal{X}^n$ a countable basis of open sets. We construct an orbit that intersects all the U_n.
By transitivity, $\exists \, n_0 \in \mathbb{N}, U_0 \cap f^{n_0}(U_1) \neq \varnothing$. Let:

- V_0 an open set such that $\overline{V_0} \subset U_0 \cap f^{-n_0}(U_1)$,

- etc.,

- V_i an open set such that $\overline{V_i} \subset U_0 \cap f^{-n_i}(U_i)$.

By compactness, $V = \bigcup_{i \in \mathbb{N}} \overline{V_i}$ is nonempty, and $\forall x \in V, \forall i \in \mathbb{N}$, $x \in f^{n_1}(x) \in U_i$.

3.3   Stability, Sensitivity, and Expansiveness

Periodic points have been regarded in Section 3.1 and stable subsets have been studied in Section 3.2, in order to determine how they can lead to unpredictability. It remains to investigate the general forms of orbits. Indeed, we want here to claim a system as stable when it is such that two closed points lead to two similar orbits. In the opposite situation, we will have an unpredictable behavior referred here as instability, sensitivity, or expansiveness.

3.3.1   Stability and instability

Let us firstly define what is a stable orbit.

Definition 36 (Stable orbit) A positive orbit γ_x is said to be *stable* if

$$\forall \varepsilon > 0, \exists \delta > 0, \forall y \in \mathcal{X}, d(x, y) < \delta \Longrightarrow \forall n \in \mathbb{N}, d(\gamma_x^n, \gamma_y^n) < \varepsilon$$

x is then called a *stable point* of f.

In other words, if y is close to x, then the orbit of y will be close to the one of x. In the neighborhood of a stable point x, all the points will evolve in a similar manner: in case of a small error on the initial condition, man can guarantee that the error between the observed phenomenon and the true theoretical evolution will remain small.

Instability is simply the negation of stability.

Definition 37 (Instability) A positive movement γ_x is *unstable* if

$$\exists \varepsilon > 0, \forall \delta > 0, \exists y \in \mathcal{X}, \exists n \in \mathbb{N}, d(x, y) < \delta \text{ and } d(\gamma_x^n, \gamma_y^n) \geqslant \varepsilon$$

A system in which all the points are stables is strongly predictable: it will be qualified as "stable." In the opposite situation, we will speak about unstable systems:

Definition 38 ((Un)stable system) A discrete dynamical system is *stable* if all its positive orbits are stable. It is said to be *unstable* if all its positive orbits are unstable.

3.3.2   Sensitivity to the initial conditions

Sensitivity to the initial conditions is stronger than instability. Here, ε does not depend on the point x under consideration:

Definition 39 A discrete dynamical system $(\mathcal{X}, f)$ is *sensitive to the initial conditions* if there exists $\varepsilon > 0$ such that

$$\forall x \in \mathcal{X}, \forall \delta > 0, \exists y \in \mathcal{X}, \exists n \in \mathbb{N}, d(x, y) < \delta \text{ and } d(\gamma_x^n, \gamma_y^n) \geqslant \varepsilon$$

ε is called the *sensitivity constant*.

A system is then sensitive if for each x, there exist arbitrarily close points of x whose respective orbits are separated from at least ε during the evolution of the system [66]. This ε is fixed; it is the same for each point x.

Remark 2 Not all the neighboring points of x are a necessarily separated ε during the evolution of the system: it is sufficient that at least one of such points exists in each open ball centered into x.

3.3.3   Expansiveness

3.3.3.1   Definition

In an expansive system, *all* errors on the initial position is amplified until reaching (at least) the constant of expansiveness ε:

Definition 40 (Expansiveness) A discrete dynamical system is *expansive* if

$$\exists \varepsilon > 0, \forall x \neq y, \exists n \in \mathbb{N}, d(\gamma_x^n, \gamma_y^n) \geqslant \varepsilon.$$

ε is called the *constant of expansiveness*.

Remark 3 Expansiveness is a kind of avalanche effect: any initial error is always magnified when iterating the system.

Expansiveness cannot define chaos alone. Indeed some linear systems are expansive. Nonlinearity is known as an important characteristic of chaotic phenomena. Additionally, an expansive system is predictable to a certain extent, as all the errors are *always* amplified to at least ε: there is no uncertainty at this level, and thus something is known of the behavior of the system.

3.3.3.2 Example

Let us introduce the following application, called *angle-doubling*:

Definition 41 (Angle-doubling) Let $\mathbb{S}^1 = \mathbb{R}/\mathbb{Z}$ be the circle group with its natural distance. The application *angle-doubling* is defined by

$$\begin{cases} f: \mathbb{S}^1 \longrightarrow \mathbb{S}^1 \\ \quad\ x \longmapsto 2x \end{cases} \tag{3.1}$$

This angle-doubling is expansive with $\varepsilon = 2$: obviously, any "error" is doubled at each iteration.

3.3.3.3 The case of perfect systems

It is possible to have expansive systems that are not sensitive to the initial conditions, as illustrated by the following example [66]:

Example 2 Let $\mathcal{X}$ be a finite space together with the trivial distance δ ($\delta(x,y) = 0$ if $x = y$, and $\delta(x,y) = 1$ else). Let $f : \mathcal{X} \to \mathcal{X}$ bijective.

- Then $(\mathcal{X}, f)$ is expansive, with a constant of expansiveness equal to 1.

- Contrarily, $(\mathcal{X}, f)$ is not sensitive to the initial conditions, because $\forall x \in \mathcal{X}, \forall \zeta < 1$, there is no y such that $\delta(x,y) < \zeta$ and $\delta\left(f^n(x), f^n(y)\right) \geqslant \varepsilon$, $\forall n, \forall \varepsilon$.

To prevent from having expansive systems that are not sensitive to their initial conditions, we can for instance suppose that $\mathcal{X}$ is perfect [66].

3.4 Chaos as Defined by Devaney (1989)

We are now able to give the most famous definition of mathematical chaos, namely the definition of Devaney [61].

Definition 42 (Chaotic discrete dynamical system) A discrete dynamical system is *chaotic according to Devaney* if it is regular and transitive.

The original definition of Devaney required the sensibility to the initial conditions. However Banks *et al.* have proven in [35] that this former definition was redundant on metric spaces:

Proposition 13 (Banks *et al.*) *If a discrete dynamical system on a metric space is chaotic according to the Definition 42, then it is sensitive to the initial conditions.*

Remark 4 Regularity and transitivity are topological properties, whereas the sensibility to the initial conditions is a metric property. Chaos, as it is defined here, is then purely a topological notion, which has important metrical consequences.

Let us quote a few words of Devaney to finish this section [61]: "A chaotic dynamical system is unpredictable because of the sensitive dependence on initial conditions. It cannot be broken down or simplified into two subsystems which do not interact because of topological transitivity. And in the midst of this random behavior, we nevertheless have an element of regularity." Thus fundamentally different behaviors are possible (periodicity, divergence, *etc.*), and they occur in an unpredictable manner.

3.5   Examples of Chaotic Systems

3.5.1   The angle-doubling

Angle-doubling is the first example of a dynamical system that is chaotic according to Devaney [61]:

Proposition 14 *Angle-doubling is chaotic as it is defined by Devaney.*

PROOF We check each of the three points of the definition:

- The angular distance between two points is doubled at each iteration of f. Then f is sensitive to the initial conditions.

- The topological transitivity comes from the fact that, even if the considered arc α is small, there will exist a k such that the image of this arc $f^k(\alpha)$ will cover the whole circle $\mathbb{S}^1$, and thus in particular any arc of $\mathbb{S}^1$.

- Periodic points are points such that $f^n(\theta) = n\theta$. Then θ is a periodic point if and only if
$$2^n\theta = \theta + 2k\pi$$
for a certain k, *i.e.*
$$\theta = \frac{2k\pi}{2^n - 1}$$

where $1 \leqslant k \leqslant 2^n$ must be an integer. So the periodic points of period n of f are the roots $2^n - 1$th of unity, which leads to the desired density.

Remark 5 The angle-doubling is one of the most classical example of sensibility to the initial conditions (an error is doubled at each iteration).

It is possible to describe the angle-doubling as iterations on the real line, *i.e.*, iterations of the following function:

$$f : \begin{array}{ccc} [0;1[& \longrightarrow & [0;1[\\ x & \longmapsto & 2x \ (\text{mod } 1) \end{array}$$

whose graph is given in Figure 3.1a. Figures 3.1b to 3.1d illustrate the sensibility when considering various closed x^0, which leads to completely different iterations.

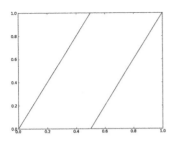

(a) The angle-doubling map.

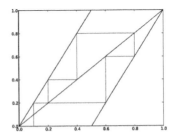

(b) Iterations for $x^0 = 0.1$.

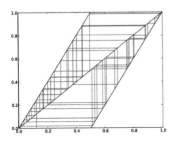

(c) Iterations for $x^0 = 0.12344$.

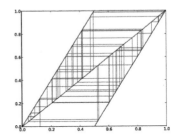

(d) Iterations for $x^0 = 0.12345$.

FIGURE 3.1: The angle-doubling.

3.5.2 The tent map

Definition 43 (Tent map) The *tent map* is defined on $[0;1]$ by

$$T(x) = \begin{cases} 2x & 0 \leqslant x \leqslant \frac{1}{2} \\ 2(1-x) & \frac{1}{2} \leqslant x \leqslant 1 \end{cases} \qquad (3.2)$$

The name of this function comes from its form (*c.f.* Figure 3.2). It is easy to show that [61]:

Proposition 15 *The dynamical system* $([0;1], T)$ *of the tent map is a chaotic system as defined by Devaney.*

A representation of the tent map and examples of iterations of its associated dynamical system are given in Figure 3.2.

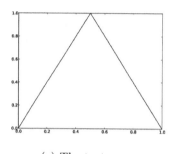

(a) The tent map.

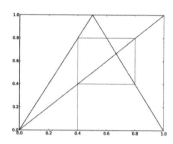

(b) Iterations for $x^0 = 0.4$.

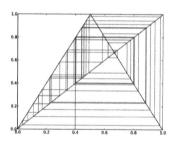

(c) Iterations for $x^0 = 0.40001$.

FIGURE 3.2: The tent map.

3.5.3   Arnold's cat map (1968)

Let us finally introduce *Arnold's cat map*[1].

Definition 44 (Arnold's cat map) *Arnold's cat map is the sequence having its values in* $[0;1]^2$, *which is defined by*

$$\begin{cases} x^{n+1} &= \quad x^n + y^n \pmod 1 \\ y^{n+1} &= \quad x^n + 2y^n \pmod 1 \end{cases}$$

---

[1]Arnold has chosen *cat* as an abbreviation of Continuous Automorphisms of the Torus.

In other words, the dynamical system consists of iterating the function $f((x,y)) = (x+y, x+2y)$ on the unity square (see Figure 3.3). It is possible to show that (the proof of this proposition is remained to the reader as an exercise):

Proposition 16 *Arnold's cat map is chaotic according to Devaney.*

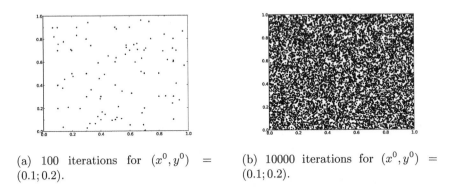

(a) 100 iterations for $(x^0, y^0) =$ (0.1; 0.2).

(b) 10000 iterations for $(x^0, y^0) =$ (0.1; 0.2).

FIGURE 3.3: Iterations of Arnold's cat map.

We will show in the state-of-the-art of the various upcoming applications that Arnold's cat map is currently one of the most used chaotic sequences (it is principally due to the fact that this is a sequence having two components).

3.6 Topological and Metrical Conjugacies

Sometimes, instead of proving directly a given property on the system itself, it is preferable to reduce the initial problem to another one whose characteristics are known or that seems to be more accessible. Such a reduction tool is called the (semi-)conjugacy in the mathematical theory of chaos.

Notions of topological and metrical conjugacy are first introduced, then properties preserved through such conjugacies are given.

3.6.1 The topological semi-conjugacy

3.6.1.1 Definition

Definition 45 (Topological semi-conjugacy) The discrete dynamical system $(\mathcal{X}, f)$ is said to be *topologically semi-conjuguate* to the system $(\mathcal{Y}, g)$ if there exists a continuous and surjective map $\varphi : \mathcal{X} \longrightarrow \mathcal{Y}$ such that:

$$\varphi \circ f = g \circ \varphi$$

that is, a map making commutative the following diagram [66]:

$$
\begin{array}{ccc}
\mathcal{X} & \xrightarrow{\ f\ } & \mathcal{X} \\
\varphi \downarrow & & \downarrow \varphi \\
\mathcal{Y} & \xrightarrow[\ g\ ]{} & \mathcal{Y}
\end{array}
$$

In that case, system $(\mathcal{Y}, g)$ is called a *factor* of the system $(\mathcal{X}, f)$.

3.6.1.2 Utility of semi-conjugacy

Several dynamical behaviors are inherited by factor systems [66]. They are summarized below:

Proposition 17 *Let $(\mathcal{Y}, g)$ a factor of the system $(\mathcal{X}, f)$. Then:*

1. *$p \in Per_k(f) \Longrightarrow \varphi(p) \in Per_j(g)$, where $j \leqslant k$,*

2. *$(\mathcal{X}, f)$ regular $\Longrightarrow (\mathcal{Y}, g)$ regular,*

3. *$(\mathcal{X}, f)$ transitive $\Longrightarrow (\mathcal{Y}, g)$ transitive,*

4. *Then: $(\mathcal{X}, f)$ chaotic as defined by Devaney $\Longrightarrow (\mathcal{Y}, g)$ chaotic as defined by Devaney.*

3.6.1.3 Example of use

Topological semi-conjugacy allows for instance to show that the logistic map is chaotic as defined by Devaney, by translating it into a new system that is better known.

Results of the Section 3.5 can thus be recovered by using the angle-doubling map:

1. $g_4(x) = 4x(1 - x)$ is linked to the angle-doubling function f in the following manner. Define $h(\theta) = (1 - cos\theta)/2$. We thus have:

$$
\begin{aligned}
(h \circ f)(q) &= \frac{1 - cos(2q)}{2} = 1 - cos^2 q \\
&= 4\left(\frac{1}{2} - \frac{cos(q)}{2}\right)\left(\frac{1}{2} + \frac{cos(q)}{2}\right) \\
&= (g_4 \circ h)(q).
\end{aligned}
$$

f is chaotic, we can thus deduce that this property can be transposed to g_4, because h is a semi-conjugacy.

2. g_4 is topologically semi-conjugate to the tent map, so this latter function is chaotic as defined by Devaney.

3.6.2 Topological conjugacy

The conjugacy (topological and metrical) is a stronger version of the semi-conjugacy.

3.6.2.1 Definitions

Topological conjugacy is defined as follows.

Definition 46 (Topological conjugacy) Two discrete dynamical systems are *topologically conjugate* if the function φ of the semi-conjugacy is a homeomorphism (bicontinuous bijection).

There exists another form of conjugacy, the metrical one:

Definition 47 (Metrical conjugacy) Two discrete dynamical systems are *metrically conjuguate* if the function φ of the semi-conjugacy is a surjective isometry.

3.6.2.2 Properties preserved by conjugacy

The stability of fixed points
When f is differentiable, its fixed points can be ordered depending on their stability:

Definition 48 (Stability and instability) A fixed point p of f is said to be *stable*, *quasi-stable*, or *unstable* according to the fact that $|f'(p)|$ is lower, equal, or greater than 1.

Depending on their stability, fixed points tend either to attract the iterates or to repel them, leading to the following definition-proposition:

Definition 49 (Attraction and repulsion) If p is stable, then it is *attractive*: there exists a neighborhood V of p such that $\forall x \in V, f^{(n)}(x) \longrightarrow p$ when $n \longrightarrow +\infty$.
If p is unstable, then it is *repellent*: there exists a neighborhood V of p whose distinct points of p leave this neighborhood during iterations of f.

Properties preserved by topological conjugacy

Proposition 18 *Let us consider the following topological conjugacy:*

$$
\begin{array}{ccc}
\mathcal{X} & \xrightarrow{\ f\ } & \mathcal{X} \\
\varphi \downarrow & & \downarrow \varphi \\
\mathcal{Y} & \xrightarrow[\ g\ ]{} & \mathcal{Y}
\end{array}
$$

Then we have:

1. *For all n, $\varphi \circ g^{(n)} = f^{(n)} \circ \varphi$,*

2. *p is a periodic point of g of period n if and only if $\varphi(p)$ is a periodic point of f of period n.*

3. *If f, g, and φ are C^1 (i.e., continuously differentiable), φ' being never equal to 0, then for p periodic point of g of period n, we have $(g^{(n)})'(p) = (f^{(n)})'(\varphi(p))$. In other words, p and $\varphi(p)$ are of the same nature.*

The conjugacy translates the fact that f and g have the same dynamics. This is a relation of equivalency:

Proposition 19 *The conjugacies, both topological and metrical, are relations of equivalency on the set of the discrete dynamical systems.*

Example 3 It is equivalent to study any of the following functions, which all belong in the quadratic family: $x \mapsto \mu x(1-x)$, $x \mapsto 1 - \lambda x^2$, $x \mapsto x^2 + c$, or $x \mapsto 2\mu x + 2x^2$... Indeed it is always possible to find an affine function that links two of these forms.

Let us finally remark that:

Proposition 20 *Stability, sensitivity, and expansiveness are metrical properties: they are preserved by any metrical conjugacy (but not by topological conjugacy).*

Chapter 4

Other Formulations of Chaos

Chaotic dynamical systems have too complex behavior to be able to encompass all of their aspects using a single "perfect" definition. Indeed there exist several notions of chaos, each of them emphasize one clearly defined aspect of a chaotic behavior. The most important definitions that remain to be introduced are presented in this section.

4.1   The Lyapunov Exponent

We first investigate notions of chaos different from the formulation of Devaney by giving a short introduction to the notion of the Lyapunov exponent.

Some dynamical systems are very sensitive to small changes in their initial condition. The constants of sensitivity to initial conditions and of expansiveness focus on that property [32]. However, these variations can quickly take on enormous proportions, growing exponentially, and neither of these two constants can measure such a behavior.

Alexander Lyapunov has examined this phenomenon and introduced an exponent that measures the rate at which these small variations can grow:

Definition 50 Let $f : \mathbb{R} \longrightarrow \mathbb{R}$ be a differentiable function. The *Lyapunov* *exponent* of the system defined by $x^0 \in \mathbb{R}$ and $x^{n+1} = f(x^n)$ is defined by

$$\lambda(x_0) = \lim_{n \to +\infty} \frac{1}{n} \sum_{i=1}^{n} \ln \left| f'\left(x^{i-1}\right) \right|.$$

Consider a dynamical system with an infinitesimal error on the initial condition x_0. When the Lyapunov exponent is positive, this error will increase (situation of chaos), whereas it will decrease if $\lambda(x_0) \leqslant 0$.

Example 4 The Lyapunov exponent of the logistic map [7] becomes positive for $\mu > 3,54$, but it is always smaller than 1. The tent map [138] and the doubling map of the circle [120] have a Lyapunov exponent both equal to $\ln(2)$.

4.2   Topological and Metrical Entropy

In this section, the definition and properties of the topological entropy are recalled.

4.2.1   Original definition of topological entropy

4.2.1.1   Introduction

The basic idea in the definition of entropy can be summarized as follows. We suppose that the initial position of the system is not known with an infinite precision, but that the behavior that will be observed by iterating the system will inform us better and better about the position of the starting point. Such assumption does not match with systems having a large topological entropy. Results will be presented in this section without demonstration, as the study of the topological entropy is not the ultimate goal of this book. The reader interested by this notion can consult, for instance, the thesis of Sylvie Ruette [123].

In 1965 Adler, Konheim, and McAndrew defined the topological entropy of a dynamical system on a compact metric space [3]; it measures the mean information quantity that an observer gains at each iteration. Intuitively, if too much information is recovered at each iteration, going largely beyond our capacity of study, then the future behavior of the system cannot be predicted. This ratio (between the quantity of information obtained from each iteration, related to our capacities of study) is an image of the notion the topological entropy attempts to formulate.

In this section, a compact metric space $(\mathcal{X}, d)$ and a continuous function $f : \mathcal{X} \to \mathcal{X}$ will be considered. The historical definition of the topological entropy will first be introduced, before giving some variants and properties that make this definition more understandable. Let us finally remark that the topological entropy is not the same notion as the Shannon entropy, as defined by the information theory (we will show in a further application chapter that this latter metric is of common use in information hiding).

4.2.1.2 Entropy of an open cover

We recall that the definition of an open cover $\mathcal{U}$ of a given set $\mathcal{X}$ has been given in Def. 13: this is an indexed family of open sets of $\mathcal{X}$ whose union is equal to $\mathcal{X}$. Let us denote by $N(\mathcal{U})$ the minimal number of opens of $\mathcal{U}$ that are needed to cover the whole set $\mathcal{X}$: if $\mathcal{U} = \{U_1; \ldots ; U_p\}$ is an open cover, then

$$N(\mathcal{U}) = \min \left\{ n \in \mathbb{N} \ / \ \exists i_1, \ldots ; i_n \in [\![1;p]\!], \mathcal{X} = U_{i_1} \cup \ldots \cup U_{i_n} \right\}.$$

Remark 6 The existence of $N(\mathcal{U})$ is ensured, as the space $\mathcal{X}$ is compact.

It remains to define a joint open cover, before being able to introduce the topological entropy of a cover:

Definition 51 (Joint open cover) Let $\mathcal{U} = \{U_1; \ldots ; U_p\}$ and $\mathcal{V} = \{V_1; \ldots ; V_q\}$ two open covers of $\mathcal{X}$. We define a new open cover, called *joint open cover* and denoted $\mathcal{U} \vee \mathcal{V}$, by intersecting these open sets two by two:

$$\mathcal{U} \vee \mathcal{V} = \left\{ U_i \cap V_j \ / \ 1 \leqslant i \leqslant p, 1 \leqslant j \leqslant q \right\}.$$

Let $N_n(\mathcal{U})$ stands for $N \left(\mathcal{U} \vee f^{-1}(\mathcal{U}) \vee \ldots \vee f^{-(n-1)}(\mathcal{U}) \right)$. It is now possible to define the entropy of a given cover.

Definition 52 (Entropy of a cover) Let $f : \mathcal{X} \to \mathcal{X}$ a continuous function, where $(\mathcal{X}, d)$ is a compact metric space. The *Entropy of the open cover* $\mathcal{U}$ of $\mathcal{X}$ is given by the formula:

$$h_{top}(\mathcal{U}, f) = \lim_{n \to +\infty} \frac{\log N_n(\mathcal{U})}{n} = \inf_{n \geqslant 1} \frac{\log N_n(\mathcal{U})}{n}.$$

4.2.1.3 The topological entropy

Definition 53 (Topological entropy) The *topological entropy* of the system $(\mathcal{X}, f)$ is defined as the supremum of the entropies of all the open covers of $\mathcal{X}$:

$$h_{top}(\mathcal{X}, f) = \sup \left\{ h_{top}(\mathcal{U}, f) \ | \ \mathcal{U} \text{ finite open cover of } \mathcal{X} \right\}.$$

Let us propose an illustration to well understand the topological notion of entropy. Suppose that positions that can take a given point x of the system under consideration are only known imperfectly. For instance, they are observed through an instrument of measure that can only return $N(\mathcal{U})$ values. The number $N_n(\mathcal{U})$ represents the minimal number of words of length n that are necessary to encode the points of $\mathcal{X}$ according to the behavior of their $n-1$ first iterates. To say it differently, the value $N_n(\mathcal{U})$ measures the number of different "scenarii" we are capable of observing for these $n-1$ iterates. In that illustration, the topological entropy is the mean quantity of information per iteration that is necessary to be able to describe the long term evolution of the system. It can be infinite.

A positive entropy implies an exponential multiplication of the number of visible orbits at each iteration. This leads to a new definition of chaos.

Definition 54 (Chaos (for topological entropy)) A dynamical system is said to be *chaotic according to the topological entropy* if its entropy is strictly positive. Chaos of the system increases with its topological entropy.

4.2.1.4   Some examples

Classical measures of topological entropy are given thereafter.

Example 5 Lipschitz continuous applications with a Lipschitz constant equal to 1 have a null entropy.

Example 6 Angle-doubling application given by Definition 41 has a topological entropy equal to $\ln 2$.

Example 7 The logistic map $f(x) = \mu x(1-x)$ of Definition 2, when $\mu > 2 + \sqrt{5}$, also has a topological entropy equal to $\ln 2$.

There is no general method used to compute the entropy of a given application. A common strategy is to transform the studied application into another function whose entropy is already known or easier to compute. Furthermore, some results allow to obtain a bound for the entropy, such as:

Proposition 21 *In case of expansive functions, the entropy is greater than the supremum of* $\dfrac{p_n}{n}$, *where p_n is equal to the number of points of period n.*

Because topological entropy is a difficult notion, a more understandable second definition will be given in the next section. This non historic definition will be used in the following sections.

4.2.2   Definition using separated sets

In this section a definition of topological entropy is introduced that is based on the separation of points, a notion that must be introduced first.

4.2.2.1   Separated points

Two points will be said to be ε-separated if they are at a distance (at least) ε one to the other, and they are ε-separated in time n, if there exists an iteration lower than n that separate them:

Definition 55 (ε-separated points) Let $\varepsilon > 0$. Two points $x, y \in \mathcal{X}$ are ε-separated if $d(x, y) > \varepsilon$.

Definition 56 (ε-separated points in time n) Let $(\varepsilon, n) \in \mathbb{R}^+ \times \mathbb{N}$. To points $x, y \in \mathcal{X}$ are ε-separated in time n if there exists $k \leqslant n$ such that $d\left(f^k(x), f^k(y)\right) > \varepsilon$.

This definition leads to a new distance, namely,

$$d_n(x, y) = \max_{0 \leqslant k \leqslant n} d\left(f^k(x), f^k(y)\right),$$

which measures the greatest distance between orbits of x and y during the n first iterations of f.

4.2.2.2 Separated sets

The (n, ε)-separated sets are sets of points that will be ε-separated in time n.

Definition 57 ((n, ε)-separated sets) A subset E of $\mathcal{X}$ is a (n, ε)-separated set if $\forall x, y \in E, x \neq y, \exists k \in [\![0; n-1]\!], d\left(f^k(x), f^k(y)\right) > \varepsilon$.

Remark 7 The time k of the real separation of two given points x and y in a (n, ε)-separated set depends on both x and y, that is, $k = k(x, y)$. Moreover it is possible that, at time $k(x, y) + 1$, these two points become not ε-separated. However, it is certain that, for all couple of points, there will be at least one separation before the iteration number n.

The maximal cardinality of a (n, ε)-separated subset of Y is denoted by $s_n(\varepsilon, Y)$. $s_n(\varepsilon, Y)$ is thus the number of points of the largest set of ε-separated points in time n of Y. Similarly, this is the set of points that will be at distance ε at a given moment during the n first iterations of f. $s_n(\varepsilon, Y)$ is thus the number of orbits of length n that can be distinguished, if we suppose that it is impossible to make the distinction between two points that are not ε-separated.

The topological entropy can be defined using this notion of separation, as follows.

4.2.2.3 Topological entropy

In [41] and [42], Bowen shows in 1971 that the topological entropy can be computed using (n, ε)-separated sets:

Proposition 22 (Bowen formula) Let $f : \mathcal{X} \longrightarrow \mathcal{X}$ a continuous function on a compact metric space $(\mathcal{X}, d)$. Then:

$$h_{top}(\mathcal{X}, f) = \lim_{\varepsilon \to 0} \left[\limsup_{n \to +\infty} \frac{1}{n} \log s_n(\varepsilon, \mathcal{X}) \right].$$

Remark 8 The function $\varepsilon \mapsto \limsup\limits_{n \to +\infty} \left(\dfrac{1}{n} \log s_n(\varepsilon, \mathcal{X}) \right)$ increases when ε decreases. This limit is thus well defined.

Thus the topological entropy measures the mean exponential growth of the number of distinguishable orbits: if the maximal number of points, whose orbits are not pasted during n iterations, are counted and if the growth speed of this quantity is measured, then we obtain a measure of the dispersion that is a characteristic of the system.

Chaos, as defined by the topological entropy, is finally somewhat equivalent to considering that a system is complex when it disperses closed states.

4.2.3   Definition using covering sets

There is another variant of the definition of the topological entropy, dual to the preceding one, which leads to a better understanding of that measure.

4.2.3.1 (n, ε)-covers

Definition 58 ((n, ε)-cover) A subset E of $\mathcal{X}$ is a (n, ε)-cover of Y if Y is into the union of open balls for the distance d_n, centered on points of E, and with a radius equal to ε:

$$Y \subset \bigcup_{x \in E} B_n(x, \varepsilon),$$

where B_n is the open ball for the distance d_n.

In other words, E is a (n, ε)-cover of Y if any point of Y will be at a distance lower than ε from at least one of the points of E during the n first iterations of the system.

Let $r_n(\varepsilon, \mathcal{X})$ denotes the minimal cardinality of a (n, ε)-cover of $\mathcal{X}$. Indeed $r_n(\varepsilon, \mathcal{X})$ is the minimal number of open balls of radius equal to ε for d_n that cover the whole $\mathcal{X}$. $r_n(\varepsilon, \mathcal{X})$ is thus the smallest number of points possessed by a cover Y, such that any point of $\mathcal{X}$ will be at distance lesser than ε of a point of Y during one of the n first iterations of the system.

4.2.3.2   Topological entropy

These cover sets lead to a new mean to define topological entropy, which has been introduced by Bowen:

Proposition 23 (Bowen formula) *Let $f : \mathcal{X} \longrightarrow \mathcal{X}$ be a continuous map on a compact metric space $\mathcal{X}$. Then*

$$h_{top}(\mathcal{X}, f) = \lim_{\varepsilon \to 0} \limsup_{n \to +\infty} \frac{1}{n} \log r_n(\varepsilon, \mathcal{X}).$$

In this variant, the goal is to measure the complexity of the dynamics of a system. Indeed, a set of initial states (*i.e.*, the cover) is a characteristic of the system, if the knowledge of the evolution of the points of this cover during the n first iterations allows one to deduce the orbit of any initial state on a similar duration.

Remark 9 If $\mathcal{X}$ is compact and $\mathcal{Y} \subset \mathcal{X}$, Y non necessarily invariant, the topological entropy of $\mathcal{Y}$ is defined in a similar manner, that is:

$$h_{top}(\mathcal{Y}, f) = \lim_{\varepsilon \to 0} \limsup_{n \to +\infty} \frac{1}{n} \log s_n(\varepsilon, \mathcal{Y}) = \lim_{\varepsilon \to 0} \limsup_{n \to +\infty} \frac{1}{n} \log r_n(\varepsilon, \mathcal{Y}).$$

Remark 10 As $\mathcal{X}$ is compact, two metrics giving the same topology are uniformly equivalent. This remark implies that the entropy defined above does not depend on the metric that has been chosen, which explains why the entropy is qualified as topological.

4.2.4 Properties of the topological entropy

This section ends with properties that are satisfied by the topological entropy. These properties will be used in a further chapter, to measure the entropy of a system relevant in the computer science security field.

Proposition 24 *Let $\mathcal{X}$ be a compact metric space and $f : \mathcal{X} \longrightarrow \mathcal{X}$ a continuous function. Then:*

- *For all integer $n \geqslant 1$, we have $h_{top}(\mathcal{X}, f^n) = n \, h_{top}(\mathcal{X})$*
- *If f is bijective, then $h_{top}(\mathcal{X}, f^{-1}) = h_{top}(\mathcal{X}, f)$*

Let us finally remark that:

Proposition 25 *The topological entropy is an invariant of the topological conjugacy.*

Part III

From theory to practice

Chapter 5

A Fundamental Tool: The Chaotic Iterations

5.1 Introducing the Chaotic Iterations

Let us consider a *system* with a finite number $N \in \mathbb{N}^*$ of elements (or *cells*), so that each cell has a Boolean *state*. Having N Boolean values for these cells leads to the definition of a particular *state of the system*. A sequence in which elements belong to $[\![1; N]\!]$ is called a *strategy*. The set of all strategies is denoted by $\mathbb{S}$.

Definition 59 Let $f : \mathbb{B}^N \longrightarrow \mathbb{B}^N$ be a function and $S \in \mathbb{S}$ be a strategy.

The so-called *chaotic iterations* are defined by $x^0 \in \mathbb{B}^N$ and

$$\forall n \in \mathbb{N}^*, \forall i \in [\![1; N]\!], x_i^n = \begin{cases} x_i^{n-1} & \text{if } S^n \neq i \\ \left(f(x^{n-1})\right)_{S^n} & \text{if } S^n = i. \end{cases}$$

In other words, at the nth iteration, only the S^n-th cell is "iterated."Note that in a more general formulation, S^n can be a subset of components and $\left(f(x^{n-1})\right)_{S^n}$ can be replaced by $\left(f(x^k)\right)_{S^n}$, where $k < n$, describing for example, delayed transmission [74, 121]. Finally, let us remark that the term "chaotic," in the name of these iterations, has *a priori* no link with the mathematical theory of chaos recalled in a previous chapter.

5.2   Chaotic Iterations as Devaney's Chaos

In this section is proven that chaotic iterations are a particular case of topological chaos, as it is defined in Devaney's formulation.

5.2.1   The new topological space

We first define a suitable metric space where chaotic iterations are continuous.

5.2.1.1   Defining the iteration function and the phase space

Let δ be the *discrete Boolean metric*, $\delta(x, y) = 0 \Leftrightarrow x = y$. Given a function f, define the function:

$$\begin{aligned} F_f : \quad [\![1; N]\!] \times \mathbb{B}^N &\longrightarrow \mathbb{B}^N \\ (k, E) &\longmapsto \left(E_j.\delta(k, j) + f(E)_k.\overline{\delta(k, j)}\right)_{j \in [\![1;N]\!]}, \end{aligned}$$

where $+$ and $.$ are the Boolean addition and product operations. Consider the phase space:

$$\mathcal{X} = [\![1; N]\!]^{\mathbb{N}} \times \mathbb{B}^N,$$

and the map defined on $\mathcal{X}$:

$$G_f (S, E) = \left(\sigma(S), F_f(i(S), E)\right), \tag{5.1}$$

where σ is the *shift* function defined by $\sigma(S^n)_{n \in \mathbb{N}} \in \mathbb{S} \longrightarrow (S^{n+1})_{n \in \mathbb{N}} \in \mathbb{S}$ and i is the *initial function* $i : (S^n)_{n \in \mathbb{N}} \in \mathbb{S} \longrightarrow S^0 \in [\![1; N]\!]$. Then the chaotic iterations defined in (5.1) can be described by the following iterations:

$$\begin{cases} X^0 \in \mathcal{X} \\ X^{k+1} = G_f(X^k). \end{cases}$$

With this formulation, a shift function appears as a component of chaotic iterations. The shift function is a famous example of a chaotic map [61] but its presence is not sufficient to claim G_f as chaotic. In the remainder of this section we rigorously prove that, under some hypotheses, chaotic iterations generate a topological chaos.

5.2.1.2 Cardinality of $\mathcal{X}$

By comparing $\mathbb{S}$ and $\mathbb{R}$, we have the following result.

Theorem 2 *The phase space $\mathcal{X}$ has, at least, the cardinality of the continuum.*

PROOF Let φ be the map which transforms a strategy into the binary representation of an element in $[0, 1[$, as follows.

- If the nth term of the strategy is 0, then the nth associated digit is 0.

- If this n^{th} term is not equal to 0, then the associated digit is 1.

With this construction, $\varphi : [\![1; N]\!]^{\mathbb{N}} \longrightarrow [0, 1]$ is onto. But $]0, 1[$ is isomorphic to $\mathbb{R}$ (indeed $x \in]0, 1[\mapsto tan(\pi(x - \frac{1}{2}))$ is an isomorphism), so the cardinality of $[\![1; N]\!]^{\mathbb{N}}$ is greater or equal to the cardinality of $\mathbb{R}$. As a consequence, the cardinality of the Cartesian product $\mathcal{X} = [\![1; N]\!]^{\mathbb{N}} \times \mathbb{B}^{\mathbb{N}}$ is greater or equal to the cardinality of $\mathbb{R}$.

Remark 11 This result is independent from the number of components of the system.

5.2.1.3 A new distance

Let us define a new distance between two points $X = (S, E), Y = (\check{S}, \check{E}) \in \mathcal{X}$ by

$$d(X, Y) = d_e(E, \check{E}) + d_s(S, \check{S}),$$

where

$$
\begin{cases}
d_e(E, \check{E}) &= \displaystyle\sum_{k=1}^{N} \delta(E_k, \check{E}_k), \\
d_s(S, \check{S}) &= \displaystyle\frac{9}{N} \sum_{k=1}^{\infty} \frac{|S^k - \check{S}^k|}{10^k}.
\end{cases}
$$

This new distance has been introduced in [32] to satisfy the following requirements.

- When the number of different cells between two systems is increasing, then their distance should increase too.

- In addition, if two systems present the same cells and their respective strategies start with the same terms, then the distance between these two points must be small because the evolution of the two systems will be the same for a while. Indeed, the two dynamical systems start with the same initial condition, use the same update function, and as strategies are the same for a while, then components that are updated are the same too.

The distance presented above follows these recommendations. Indeed, if the floor value $\lfloor d(X,Y) \rfloor$ is equal to n, then the systems $E, \check{E}$ differ in n cells. In addition, $d(X,Y) - \lfloor d(X,Y) \rfloor$ is a measure of the differences between strategies S and $\check{S}$. More precisely, this floating part is less than 10^{-k} if and only if the first k terms of the two strategies are equal. Moreover, if the kth digit is nonzero, then the kth terms of the two strategies are different.

5.2.1.4   Continuity of the iteration function

To prove that chaotic iterations are an example of topological chaos in the sense of Devaney [61], G_f must be continuous in the metric space $(\mathcal{X}, d)$.

Theorem 3 G_f *is a continuous function.*

PROOF We use the sequential continuity. Let $(S^n, E^n)_{n \in \mathbb{N}}$ be a sequence of the phase space $\mathcal{X}$, which converges to (S,E). We will prove that $\big(G_f(S^n, E^n)\big)_{n \in \mathbb{N}}$ converges to $\big(G_f(S,E)\big)$. Let us recall that for all n, S^n is a strategy, thus, we consider a sequence of strategies (*i.e.*, a sequence of sequences).

As $d((S^n, E^n); (S,E))$ converges to 0, each distance $d_e(E^n, E)$ and $d_s(S^n, S)$ converges to 0. But $d_e(E^n, E)$ is an integer, so $\exists n_0 \in \mathbb{N}, d_e(E^n, E) = 0$ for any $n \geqslant n_0$.
In other words, there exists a threshold $n_0 \in \mathbb{N}$ after which no cell will change its state: $\exists n_0 \in \mathbb{N}, n \geqslant n_0 \Rightarrow E^n = E$.
In addition, $d_s(S^n, S) \longrightarrow 0$, so $\exists n_1 \in \mathbb{N}, d_s(S^n, S) < 10^{-1}$ for all indices greater than or equal to n_1. This means that for $n \geqslant n_1$, all the S^n have the same first term, which is S^0: $\forall n \geqslant n_1, S_0^n = S_0$.
Thus, after the $max(n_0, n_1)^{th}$ term, states of E^n and E are identical and strategies S^n and S start with the same first term.
Consequently, states of $G_f(S^n, E^n)$ and $G_f(S,E)$ are equal, so, after the $max(n_0, n_1)^{th}$ term, the distance d between these two points is strictly less than 1.
We now prove that the distance between $\big(G_f(S^n, E^n)\big)$ and $\big(G_f(S,E)\big)$ is convergent to 0. Let $\varepsilon > 0$.

- If $\varepsilon \geqslant 1$, we see that distance between $\big(G_f(S^n, E^n)\big)$ and $\big(G_f(S,E)\big)$ is strictly less than 1 after the $max(n_0, n_1)^{th}$ term (same state).

- If $\varepsilon < 1$, then $\exists k \in \mathbb{N}, 10^{-k} \geqslant \varepsilon > 10^{-(k+1)}$. But $d_s(S^n, S)$ converges to 0, so
$$\exists n_2 \in \mathbb{N}, \forall n \geqslant n_2, d_s(S^n, S) < 10^{-(k+2)},$$
thus after n_2, the $k + 2$ first terms of S^n and S are equal.

As a consequence, the $k + 1$ first entries of the strategies of $G_f(S^n, E^n)$ and $G_f(S, E)$ are the same (G_f is a shift of strategies) and due to the definition of d_s, the floating part of the distance between (S^n, E^n) and (S, E) is strictly less than $10^{-(k+1)} \leqslant \varepsilon$.

In conclusion,

$$\forall \varepsilon > 0, \exists N_0 = max(n_0, n_1, n_2) \in \mathbb{N}, \forall n \geqslant N_0, d\left(G_f(S^n, E^n); G_f(S, E)\right) \leqslant \varepsilon.$$

G_f is consequently continuous.

In this section, we proved that chaotic iterations can be modeled as a dynamical system in a topological space. In the next section, we show that some chaotic iterations behave chaotically, as defined by Devaney's theory.

5.2.2 Discrete chaotic iterations as topological chaos

To prove that we are in the framework of Devaney's topological chaos, we have to find a Boolean function f such that G_f satisfies the regularity, transitivity, and sensitivity conditions. We will prove that the vectorial logical negation

$$f_0(x_1, \ldots, x_N) = (\overline{x_1}, \ldots, \overline{x_N}) \tag{5.2}$$

is a suitable function.

5.2.2.1 Regularity

First, let us prove that,

Theorem 4 *Periodic points of G_{f_0} are dense in $\mathcal{X}$.*

PROOF Let $(\check{S}, \check{E}) \in \mathcal{X}$ and $\varepsilon > 0$. We are looking for a periodic point $(\widetilde{S}, \widetilde{E})$ satisfying $d((\check{S}, \check{E}); (\widetilde{S}, \widetilde{E})) < \varepsilon$. As ε can be strictly less than 1, we must choose $\widetilde{E} = \check{E}$. Let us define $k_0(\varepsilon) = \lfloor log_{10}(\varepsilon) \rfloor + 1$ and consider the set

$$\mathcal{S}_{\check{S}, k_0(\varepsilon)} = \left\{ S \in \mathbb{S}/S^k = \check{S}^k, \forall k \leqslant k_0(\varepsilon) \right\}.$$

Then, $\forall S \in \mathcal{S}_{\check{S}, k_0(\varepsilon)}, d((S, \check{E}); (\check{S}, \check{E})) < \varepsilon$. It remains to choose $\widetilde{S} \in \mathcal{S}_{\check{S}, k_0(\varepsilon)}$ such that $(\widetilde{S}, \widetilde{E}) = (\widetilde{S}, \check{E})$ is a periodic point for G_{f_0}. Let

$$\mathcal{J} = \left\{ i \in \{1, ..., \mathsf{N}\}/E_i \neq \check{E}_i, \text{ where } (S, E) = G_{f_0}^{k_0}(\check{S}, \check{E}) \right\},$$

$i_0 = card(\mathcal{J})$, and $j_1 < j_2 < ... < j_{i_0}$ the elements of $\mathcal{J}$. Then, $\widetilde{S} \in \mathcal{S}_{\check{S}, k_0(\varepsilon)}$ defined by

- $\widetilde{S}^k = \check{S}^k$, if $k \leqslant k_0(\varepsilon)$,

- $\widetilde{S}^k = j_{k-k_0(\varepsilon)}$, if $k \in \{k_0(\varepsilon) + 1, k_0(\varepsilon) + 2, ..., k_0(\varepsilon) + i_0\}$,

- and $\widetilde{S}^k = \widetilde{S}^j$, where $j \leqslant k_0(\varepsilon) + i_0$ is satisfying $j \equiv k \pmod{k_0(\varepsilon) + i_0}$, if $k > k_0(\varepsilon) + i_0$,

is such that $(\widetilde{S}, \widetilde{E})$ is a periodic point (of period $k_0(\varepsilon) + i_0$), which is ε-close to $(\check{S}, \check{E})$.

As a conclusion, $(\mathcal{X}, G_{f_0})$ is regular.

5.2.2.2 Transitivity

Regarding the transitivity property of G_{f_0}, we can show that,

Theorem 5 $(\mathcal{X}, G_{f_0})$ *is topologically transitive.*

PROOF Let us define $\mathcal{E} : \mathcal{X} \to \mathbb{B}^{\mathbb{N}}$, such that $\mathcal{E}(S, E) = E$. Let $\mathcal{B}_A = \mathcal{B}(X_A, r_A)$ and $\mathcal{B}_B = \mathcal{B}(X_B, r_B)$ be two open balls of $\mathcal{X}$, with $X_A = (S_A, E_A)$ and $X_B = (S_B, E_B)$. We are looking for $\widetilde{X} = (\widetilde{S}, \widetilde{E})$ in $\mathcal{B}_A$ such that $\exists n_0 \in \mathbb{N}, G_{f_0}^{n_0}(\widetilde{X}) \in \mathcal{B}_B$.

$\widetilde{X}$ must be in $\mathcal{B}_A$ and r_A can be strictly lesser than 1, so $\widetilde{E} = E_A$. Let $k_0 = \lfloor \log_{10}(r_A) + 1 \rfloor$. Then $\forall S \in \mathbb{S}$, if $S^k = S_A^k, \forall k \leqslant k_0$, then $(S, \widetilde{E}) \in \mathcal{B}_A$. Let $(\check{S}, \check{E})$ be equal to $G_{f_0}^{k_0}(S_A, E_A)$ and $c_1, ..., c_{k_1}$ denote the elements of the set $\{i \in [\![1, \mathbb{N}]\!]/\check{E}_i \neq \mathcal{E}(X_B)_i\}$. So any point X of the set

$$\{(S, E_A) \in \mathcal{X} / \forall k \leqslant k_0, S^k = S_A^k \text{ and } \forall k \in [\![1, k_1]\!], S^{k_0+k} = c_k\}$$

is satisfying $X \in \mathcal{B}_A$ and $\mathcal{E}\left(G_{f_0}^{k_0+k_1}(X)\right) = E_B$. Lastly, let k_2 be $\lfloor \log_{10}(r_B) \rfloor + 1$. Then $\widetilde{X} = (\widetilde{S}, \widetilde{E}) \in \mathcal{X}$ defined by:

1. $\widetilde{X} = E_A$,

2. $\forall k \leqslant k_0, \widetilde{S}^k = S_A^k$,

3. $\forall k \in [\![1, k_1]\!], \widetilde{S}^{k_0+k} = c_k$,

4. $\forall k \in \mathbb{N}^*, \widetilde{S}^{k_0+k_1+k} = S_B^k$,

is such that $\widetilde{X} \in \mathcal{B}_A$ and $G_{f_0}^{k_0+k_1}(\widetilde{X}) \in \mathcal{B}_B$. This fact concludes the proof of the theorem.

5.2.2.3 Devaney's chaos

In conclusion, $(\mathcal{X}, G_{f_0})$ is topologically transitive and regular. Then we have the following result:

Theorem 6 G_{f_0} *is a chaotic map on* $(\mathcal{X}, d)$ *in the sense of Devaney.*

We have proven that the set $\mathcal{C}$ of the iterate functions f so that $(\mathcal{X}, G_f)$ is chaotic (according to the definition of Devaney), is a nonempty set. In a future section, we will deepen the study of $\mathcal{C}$, among other things, by computing its cardinality and characterizing this set.

5.3 Topological Properties of Chaotic Iterations

In this section, some qualitative and quantitative topological properties for chaotic iterations with G_{f_0} will be studied in detail. These properties reinforce the chaotic behavior of the system.

5.3.1 Topological mixing

We have the result [78],

Theorem 7 $(\mathcal{X}, G_{f_0})$ *is topologically mixing.*

This result is an immediate consequence of the lemma below.

Lemma 1 *For any open ball* B *of* $\mathcal{X}$, *an index* n *can be found such that* $G_{f_0}^n(B) = \mathcal{X}$.

PROOF Let $B = B((E, S), \varepsilon)$ be an open ball, of which the radius can be considered as strictly less than 1. The elements of B all have the same state E and are such that an integer $k \left(= -\log_{10}(\varepsilon)\right)$ satisfies:

- all the strategies of B have the same k first terms,

- after the index k, all values are possible.

Then, after k iterations, the new state of the system is $G_{f_0}^k(E, S)_1$ and all the strategies are possibles (any point of the form $(G_{f_0}^k(E, S)_1, \hat{S})$, with any $\hat{S} \in \mathbb{S}$, is reachable from B).

Let $(E', S') \in \mathcal{X}$. We will prove that any point of $\mathcal{X}$ is reachable from B.

Indeed, let s_i be the list of the different cells between $G_{f_0}^k(E, S)_1$ and E', and $|s|$ its size. The point $(\check{E}, \check{S})$ of B defined by:

- $\check{E} = E$,

- $\check{S}^i = S^i, \forall i \leqslant k,$

- $\check{S}^{k+i} = s_i, \forall i \leqslant |s|,$

- $\forall i \in \mathbb{N}, S^{k+|s|+i} = S'^i.$

is such that $G_{f_0}^{k+|s|}(\check{E}, \check{S}) = (E', S')$. This concludes the proof of the lemma.

5.3.2   Quantitative measures

In Section 5.2.2.3 we have proven that discrete chaotic iterations produce a topological chaos by checking two qualitative properties, namely, transitivity and regularity. As stated in the second part of this book, the mathematical theory of chaos offers tools to measure this chaos quantitatively. They are investigated in what follows.

5.3.2.1   Sensitivity

The sensitive dependence on the initial conditions has been shown as a consequence of the regularity and the transitivity of chaotic iterations. However, in the set of machine numbers, we have shown in [32] that the notion of regularity must be redefined. This is the reason why this sensitivity should be proven without using the result of Banks [35], to be sure that this dependence is preserved in practical use of chaotic iterations.

In addition, the constant of sensitivity will be obtained during this proof.

Theorem 8 $(\mathcal{X}, G_{f_0})$ *has sensitive dependence on initial conditions and its constant of sensitivity is equal to* $\mathsf{N} - 1$.

PROOF Let $\check{X} = (\check{S}, \check{E}) \in \mathcal{X}$. We are looking for $\widetilde{X} = (\widetilde{S}, \widetilde{E}) \in \mathcal{X}$ such that $d(\check{X}, \widetilde{X}) \leqslant \delta$ and $\exists n_0 \in \mathbb{N}, d\left(G_{f_0}^{n_0}(\check{X}); G_{f_0}^{n_0}(\widetilde{X})\right) \geqslant \mathsf{N} - 1$. Let k_0 be $\lfloor \log_{10}(\delta) \rfloor + 1$. So, if $S \in \{S \in \mathbb{S}/\forall k \leqslant k_0, S^k = \check{S}^k\}$, then $d\left((S, \check{E}), (\check{S}, \check{E})\right) \leqslant \delta$.

Let $\mathcal{J} = \left\{i \in \llbracket 1, \mathsf{N} \rrbracket / \; \mathcal{E}\left(G_{f_0}^{k_0}(\check{S}, \check{E})\right)_i = \mathcal{E}\left(G_{f_0}^{k_0+\mathsf{N}}(\check{S}, \check{E})\right)_i\right\}$ and $p = card\,(\mathcal{J})$. If $p = \mathsf{N}$, then $(\widetilde{S}, \widetilde{E}) \in \mathcal{X}$, defined by

1. $\widetilde{E} = \check{E}$,

2. $\forall k \leqslant k_0, \widetilde{S}^k = \check{S}^k$,

3. $\forall k \in \llbracket 1, \mathsf{N} \rrbracket, \widetilde{S}^{k_0+k} = k$,

4. $\forall k > k_0 + \mathsf{N}, \widetilde{S}^k = 1$.

satisfies $d((\widetilde{S}, \widetilde{E}); (\check{S}, \check{E})) < \delta$ and $\forall i \in \llbracket 1, \mathsf{N} \rrbracket, \mathcal{E}\left(G_{f_0}^{k_0+\mathsf{N}}(\widetilde{S}; \widetilde{E})\right)_i \neq \mathcal{E}\left(G_{f_0}^{k_0+\mathsf{N}}(\check{S}; \check{E})\right)_i$, so the result is obtained.

Else, let $j_1 < j_2 < ... < j_p$ be the elements of $\mathcal{J}$ and $j_0 \notin \mathcal{J}$. Then $\widetilde{X} = (\widetilde{E}, \widetilde{S}) \in \mathcal{X}$ defined by

1. $\widetilde{E} = \check{E}$,

2. $\forall k \leqslant k_0, \widetilde{S}^k = \check{S}^k$,

3. $\forall k \in [\![1, p]\!], \widetilde{S}^{k_0+k} = j_k$,

4. $\forall k \in \mathbb{N}^*, \widetilde{S}^{k_0+p+k} = j_0$.

is such that $d(\check{X}, \widetilde{X}) < \delta$. In addition, $\forall i \in [\![1, p]\!], \mathcal{E}\left(G_{f_0}^{k_0+\mathsf{N}}(\check{X})\right)_{j_i} \neq \mathcal{E}\left(G_{f_0}^{k_0+\mathsf{N}}(\widetilde{X})\right)_{j_i}$, because:

- $\forall i \in [\![1, \mathsf{N}]\!], \mathcal{E}\left(G_{f_0}^{k_0}(\check{X})\right)_i = \mathcal{E}\left(G_{f_0}^{k_0}(\widetilde{X})\right)_i$, due to the definition of k_0.

- $\forall i \in [\![1, p]\!], j_i \in \mathcal{J} \Rightarrow \mathcal{E}\left(G_{f_0}^{k_0+\mathsf{N}}(\check{X})\right)_{j_i} = \mathcal{E}\left(G_{f_0}^{k_0}(\check{X})\right)_{j_i}$, according to the definition of $\mathcal{J}$.

- $\forall i \in [\![1, p]\!], j_i$ appears exactly one time in $\widetilde{S}^{k_0}, \widetilde{S}^{k_0+1}, ..., \widetilde{S}^{k_0+\mathsf{N}}$, so

$$\mathcal{E}\left(G_{f_0}^{k_0+\mathsf{N}}(\widetilde{X})\right)_{j_i} \neq \mathcal{E}\left(G_{f_0}^{k_0}(\widetilde{X})\right)_{j_i}.$$

Last, $\forall i \in [\![1, \mathsf{N}]\!] \setminus \{j_0, j_1, ..., j_p\}, \mathcal{E}\left(G_{f_0}^{k_0+\mathsf{N}}(\widetilde{X})\right)_i \neq \mathcal{E}\left(G_{f_0}^{k_0+\mathsf{N}}(\check{X})\right)_i$, because:

- $\forall i \in [\![1, \mathsf{N}]\!], \mathcal{E}\left(G_{f_0}^{k_0}(\check{X})\right)_i = \mathcal{E}\left(G_{f_0}^{k_0}(\widetilde{X})\right)_i$,

- $i \notin \mathcal{J} \Rightarrow \mathcal{E}\left(G_{f_0}^{k_0+\mathsf{N}}(\check{X})\right)_i \neq \mathcal{E}\left(G_{f_0}^{k_0}(\check{X})\right)_i$,

- $i \notin \{\widetilde{S}^{k_0}, \widetilde{S}^{k_0+1}, ..., \widetilde{S}^{k_0+\mathsf{N}}\} \Rightarrow \mathcal{E}\left(G_{f_0}^{k_0+\mathsf{N}}(\widetilde{X})\right)_i = \mathcal{E}\left(G_{f_0}^{k_0}(\widetilde{X})\right)_i$.

So, in this case, $\forall i \in [\![1, \mathsf{N}]\!] \setminus \{j_0\}, \mathcal{E}\left(G_{f_0}^{k_0+\mathsf{N}}(\widetilde{S}; \widetilde{E})\right)_i \neq \mathcal{E}\left(G_{f_0}^{k_0+\mathsf{N}}(\check{S}; \check{E})\right)_i$ and the result of sensitivity is still obtained.

5.3.2.2 Expansiveness

In this section we offer the proof that chaotic iterations are expansive [78] when f_0 is the update function:

Theorem 9 $(\mathcal{X}, G_{f_0})$ *is an expansive chaotic system. Its constant of expansiveness is equal to 1.*

PROOF If $(S, E) \neq (\check{S}; \check{E})$, then:

- Either $E \neq \check{E}$ and so at least one cell is not in the same state in E and $\check{E}$. Consequently the distance between (S, E) and $(\check{S}; \check{E})$ is greater or equal to 1.

- Or $E = \check{E}$. So the strategies S and $\check{S}$ are not equal. Let n_0 be the first index such that the terms S and $\check{S}$ differ. Then $\forall k < n_0, G_{f_0}^{n_0}(S, E) = G_{f_0}^k(\check{S}, \check{E})$, and $G_{f_0}^{n_0}(S, E) \neq G_{f_0}^{n_0}(\check{S}, \check{E})$.

  As $E = \check{E}$, the cell that has changed in E at the n_0-th iterate is not the same as the cell that has changed in $\check{E}$, so the distance between $G_{f_0}^{n_0}(S, E)$ and $G_{f_0}^{n_0}(\check{S}, \check{E})$ is greater or equal to 2.

So the expansiveness property is established.

Remark 12 $(\mathcal{X}, G_{f_0})$ is not A-expansive, for any $A > 1$: let us consider two points $X = (E, S)$ and $X' = (E', S')$ with the same strategy $(S = S')$ and only one different cell $(d_e(E, E') = 1)$. So, $\forall n \in \mathbb{N}, d_e\left(\mathcal{E}\left(G_{f_0}^n(X)\right), \mathcal{E}\left(G_{f_0}^n(X')\right)\right) = 1$.

5.3.3 Topological Entropy

As detailed in Chapter 4, another important tool to measure the chaotic behavior of a dynamical system is the topological entropy, which is defined for compact topological spaces. Before studying the entropy of CIs, we must then check that $(\mathcal{X}, d)$ is compact.

5.3.3.1 Compactness study

In this section, we will prove that $(\mathcal{X}, d)$ is a compact topological space, in order to study its topological entropy later. First, as $(\mathcal{X}, d)$ is a metric space, it is separated. It is, however, possible to give a direct proof of this result:

Theorem 10 $(\mathcal{X}, d)$ *is a separated space.*

PROOF Let $(E, S) \neq (\hat{E}, \hat{S})$ two points of $\mathcal{X}$.

1. If $E \neq \hat{E}$, then the intersection between the two balls $\mathcal{B}\left((E, S), \frac{1}{2}\right)$ and $\mathcal{B}\left((\hat{E}, \hat{S}), \frac{1}{2}\right)$ is empty.

2. Or it exists $k \in \mathbb{N}$ such that $S^k \neq \hat{S}^k$ then the balls $\mathcal{B}\left((E, S), 10^{-(k+1)}\right)$ and $\mathcal{B}\left((\hat{E}, \hat{S}), 10^{-(k+1)}\right)$ can be chosen.

The sequential characterization of the compacity for metric spaces can now be used to obtain the following result.

Theorem 11 $(\mathcal{X}, d)$ *is a compact metric space.*

PROOF Let $(E^n, S^n)_{n \in \mathbb{N}}$ be a sequence of $\mathcal{X}$.

1. A state $E^{\tilde{n}}$ which appears an infinite number of time in this sequence can be found. Let
$$I = \{(E^n, S^n)/E^n = E^{\tilde{n}}\}.$$
For all $(E, S) \in I$, $S_0^n \in [\![1, N]\!]$, and I is an infinite set. Then $\tilde{k} \in [\![1, N]\!]$ can be found such that an infinite number of strategies of I start with $\tilde{k}$.

   Let n_0 be the smallest integer such that $E^n = E^{\tilde{n}}$ and $S_0^n = \tilde{k}$.

2. The set
$$I' = \{(E^n, S^n)/E^n = E^{n_0} \text{ and } S_0^n = S_0^{n_0}\}$$
is infinite; one of the elements of $[\![1, N]\!]$ will appear an infinite number of times in the S_1^n of I': let us call it $\tilde{1}$.

   Let n_1 be the smallest n such that $(E^n, S^n) \in I'$ and $S_1^n = \tilde{1}$.

3. The set
$$I'' = \{(E^n, S^n)|E^n = E^{n_0} \text{ and } S_0^n = S_0^{n_0} \text{ and } S_1^n = S_1^{n_1}\}$$
is infinite, *etc.*

Let $l = \left(E^{n_0}, \left(S_k^{n_k}\right)_{k \in \mathbb{N}}\right)$, then the subsequence (E^{n_k}, S^{n_k}) converges to l.

5.3.3.2 Topological entropy

We have the result,

Theorem 12 *The topological entropy of $(\mathcal{X}, G_f)$ is infinite.*

PROOF Let $E, \check{E} \in \mathbb{B}^N$ such that $\exists i_0 \in [\![1, N]\!], E_{i_0} \neq \check{E}_{i_0}$. Then, $\forall S, \check{S} \in \mathcal{S}$,
$$d((E, S); (\check{E}, \check{S})) \geqslant 1$$
But the cardinal c of $\mathcal{S}$ is infinite, then $\forall n \in \mathbb{N}, c > e^{n^2}$.

Then for all $n \in \mathbb{N}$, the maximal number $H(n, 1)$ of $(n, 1)$-separated points is greater than or equal to e^{n^2}, so
$$h_{top}(G_f, 1) = \overline{lim} \frac{1}{n} log\left(H(n, 1)\right) > \overline{lim} \frac{1}{n} log\left(e^{n^2}\right) = \overline{lim}\ (n) = +\infty.$$
But $h_{top}(G_f, \varepsilon)$ is an increasing function when ε is decreasing, then
$$h_{top}\left(G_f\right) = \lim_{\varepsilon \to 0} h_{top}(G_f, \varepsilon) > h_{top}(G_f, 1) = +\infty,$$
which concludes the evaluation of the topological entropy of G_f.

We have proven that it is possible to find f, such that chaotic iterations generated by f can be described by a chaotic and entropic map on a topological space in the sense of Devaney. We have considered a finite set of states $\mathbb{B}^N$ and a set $\mathbb{S}$ of strategies composed by an infinite number of infinite sequences. Next section gives a characterization of such functions.

5.4   Characterization

We now propose a characterization of Boolean networks f making the iterations of any induced map G_f chaotic [9]. This is achieved by establishing inclusion relations between the transitive, regular, and chaotic sets defined below:

- $\mathcal{T} = \left\{ f : \mathbb{B}^n \to \mathbb{B}^n / G_f \text{ is transitive} \right\}$,

- $\mathcal{R} = \left\{ f : \mathbb{B}^n \to \mathbb{B}^n / G_f \text{ is regular} \right\}$,

- $\mathcal{C} = \left\{ f : \mathbb{B}^n \to \mathbb{B}^n / G_f \text{ is chaotic (Devaney)} \right\}$.

Let f be a map from $\mathbb{B}^n$ to itself. The *asynchronous iteration graph* associated with f is the directed graph $\Gamma(f)$ defined by: the set of vertices is $\mathbb{B}^n$; for all $x \in \mathbb{B}^n$ and $i \in [\![1; n]\!]$, the graph $\Gamma(f)$ contains an arc from x to $F_f(i, x)$. The relation between $\Gamma(f)$ and G_f is clear: there exists a path from x to x' in $\Gamma(f)$ if and only if there exists a strategy s such that the parallel iteration $x^{n+1} = G_f(x^n)$ of G_f from the initial point $x^0 = (s, x)$ reaches the point x'. Finally, in what follows, the term *iteration graph* is a shortcut for asynchronous iteration graph.

We can thus characterize $\mathcal{T}$:

Proposition 26 *G_f is transitive if and only if $\Gamma(f)$ is strongly connected.*

PROOF $\Longleftarrow$ Suppose that $\Gamma(f)$ is strongly connected. Let (s, x) and (s', x') be two points of $\mathcal{X}$, and let $\varepsilon > 0$. We will define a strategy $\tilde{s}$ such that the distance between $(\tilde{s}, x)$ and (s, x) is less than ε, and such that the parallel iterations of G_f from $(\tilde{s}, x)$ reaches the point (s', x').

Let $t_1 = \lfloor -\log_{10}(\varepsilon) \rfloor$, and let x'' be the configuration of $\mathbb{B}^n$ that we obtain from (s, x) after t_1 iterations of G_f. Since $\Gamma(f)$ is strongly connected, there exists a strategy s'' and $t_2 \in \mathbb{N}$ such that, x' is reached from (s'', x'') after t_2 iterations of G_f.

Now, consider the strategy $\tilde{s} = (s_0, \ldots, s_{t_1-1}, s_0'', \ldots, s_{t_2-1}'', s_0', s_1', s_2', s_3' \ldots)$. It is clear that (s', x') is reached from $(\tilde{s}, x)$ after $t_1 + t_2$ iterations of G_f, and since $\tilde{s}_t = s_t$ for $t < t_1$, by the choice of t_1, we have $d((s, x), (\tilde{s}, x)) < \varepsilon$. Consequently, G_f is transitive.

$\Longrightarrow$ If $\Gamma(f)$ is not strongly connected, then there exist two configurations x and x' such that $\Gamma(f)$ has no path from x to x'. Let s and s' be two strategies, and let $0 < \varepsilon < 1$. Then, for all (s'', x'') such that $d((s'', x''), (s, x)) < \varepsilon$, we have $x'' = x$, so that iteration of G_f from (s'', x'') only reaches points in $\mathcal{X}$ that are at a greater distance than one with (s', x'). So G_f is not transitive.

We now prove that:

Proposition 27 $\mathcal{T} \subset \mathcal{R}$.

PROOF Let $f : \mathbb{B}^n \to \mathbb{B}^n$ such that G_f is transitive (f is in $\mathcal{T}$). Let $(s, x) \in \mathcal{X}$ and $\varepsilon > 0$. To prove that f is in $\mathcal{R}$, it is sufficient to prove that there exists a strategy $\tilde{s}$ such that the distance between $(\tilde{s}, x)$ and (s, x) is less than ε, and such that $(\tilde{s}, x)$ is a periodic point.

Let $t_1 = \lfloor - \log_{10}(\varepsilon) \rfloor$, and let x' be the configuration that we obtain from (s, x) after t_1 iterations of G_f. According to the previous proposition, $\Gamma(f)$ is strongly connected. Thus, there exists a strategy s' and $t_2 \in \mathbb{N}$ such that x is reached from (s', x') after t_2 iterations of G_f.

Consider the strategy $\tilde{s}$ that alternates the first t_1 terms of s and the first t_2 terms of s': $\tilde{s} = (s_0, \ldots, s_{t_1-1}, s'_0, \ldots, s'_{t_2-1}, s_0, \ldots, s_{t_1-1}, s'_0, \ldots, s'_{t_2-1}, s_0, \ldots)$. It is clear that $(\tilde{s}, x)$ is obtained from $(\tilde{s}, x)$ after $t_1 + t_2$ iterations of G_f. So $(\tilde{s}, x)$ is a periodic point. Since $\tilde{s}_t = s_t$ for $t < t_1$, by the choice of t_1, we have $d((s, x), (\tilde{s}, x)) < \epsilon$.

Remark 13 Inclusion of proposition 27 is strict, due to the identity map (which is regular, but not transitive).

We can thus conclude that $\mathcal{C} = \mathcal{R} \cap \mathcal{T} = \mathcal{T}$, which leads to the following characterization:

Theorem 13 *Let $f : \mathbb{B}^n \to \mathbb{B}^n$. G_f is chaotic (according to Devaney) if and only if $\Gamma(f)$ is strongly connected.*

We have achieved the characterization of $\mathcal{C}$. Let us now determine the size of $\mathcal{C}$, *i.e.*, the number of different graphs $\Gamma(f)$.

We can compute the number of graphs $\Gamma(f)$ due to the fact that these graphs satisfy the following property. For each vertex $x \in [\![0; 2^{n-1}]\!]$ and each arc $\alpha_k, k \in [\![1; n]\!]$, corresponding to the activation of the element number k, there are two possibilities:

- The element number k of the system x does not change when the strategy is equal to k (α_k is then a loop on x).

- The element number k changes its configuration, α_k is then an arc from $x = (x_1, \ldots, x_{k-1}, x_k, x_{k+1}, \ldots, x_n)$ to $(x_1, \ldots, x_{k-1}, \overline{x_k}, x_{k+1}, \ldots, x_n)$.

We have then two possibilities for each of the n arcs of the 2^n vertices. Then...

Theorem 14 *Let n be the number of elements of the system. Then there are $(2^n)^{2^n}$ different graphs $\Gamma(f)$.*

Theorem 14 means that if the system has n elements, then there are only $(2^n)^{2^n}$ "types of future evolutions," *i.e.*, which are fundamentally different.

5.5   The Lyapunov Exponent

We will now investigate the disorder generated by chaotic iterations G_{f_0} using the Lyapunov exponent.

5.5.1   The phase space is an interval of the real line

5.5.1.1   Toward a topological semiconjugacy

We show, by using a topological semiconjugacy, that chaotic iterations on $\mathcal{X}$ can be described as iterations on a real interval. To do so, some notations and terminologies must be introduced.

Let $\mathcal{S}_N = [\![1; N]\!]^{\mathbb{N}}$ be the set of sequences belonging in $[\![1; N]\!]$ and $\mathcal{X}_N = \mathcal{S}_N \times \mathbb{B}^N$. In what follows and for easy understanding, we will assume that $N = 10$. However, an equivalent formulation of the following can be easily obtained by replacing the base 10 by any base N.

Definition 60 The function $\varphi : \mathcal{S}_{10} \times \mathbb{B}^{10} \to [0, 2^{10}[$ is defined by:

$$\varphi : \quad \begin{array}{ccc} \mathcal{X}_{10} = \mathcal{S}_{10} \times \mathbb{B}^{10} & \longrightarrow & [0, 2^{10}[\\ ((S_0, S_1, \ldots); (E_0, \ldots, E_9)) & \longmapsto & \varphi((S, E)) \end{array}$$

where $(S, E) = ((S_0, S_1, \ldots); (E_0, \ldots, E_9))$, and $\varphi((S, E))$ is the real number:

- whose integral part e is $\sum\limits_{k=0}^{9} 2^{9-k} E_k$, that is, the binary digits of e are $E_0 \ E_1 \ \ldots \ E_9$.

- whose decimal part s is equal to $s = 0, S_0 \ S_1 \ S_2 \ \ldots = \sum_{k=1}^{+\infty} 10^{-k} S^{k-1}$.

φ realizes the association between a point of $\mathcal{X}_{10}$ and a real number into $[0, 2^{10}[$. We must now translate chaotic iterations G_{f_0} on this real interval. To do so, two intermediate functions over $[0, 2^{10}[$ must be introduced:

Definition 61 Let $x \in [0, 2^{10}[$ and:

- $e_0, \ldots, e_9$ the binary digits of the integral part of x: $\lfloor x \rfloor = \sum\limits_{k=0}^{9} 2^{9-k} e_k$.

- $(s_k)_{k \in \mathbb{N}}$ the digits of x, where the chosen decimal decomposition of x is the one that does not have an infinite number of 9:
$$x = \lfloor x \rfloor + \sum_{k=0}^{+\infty} s_k 10^{-k-1}.$$

e and s are thus defined as follows:

$$e: \quad \begin{array}{ccc} [0,2^{10}[& \longrightarrow & \mathbb{B}^{10} \\ x & \longmapsto & (e_0,\ldots,e_9) \end{array}$$

and

$$s: \quad \begin{array}{ccc} [0,2^{10}[& \longrightarrow & [\![0,9]\!]^{\mathbb{N}}. \\ x & \longmapsto & (s_k)_{k\in\mathbb{N}} \end{array}$$

We are now able to define the function g, whose goal is to translate the chaotic iterations G_{f_0} on an interval of $\mathbb{R}$.

Definition 62 $g: [0,2^{10}[\longrightarrow [0,2^{10}[$ is defined by:

$$g: \quad \begin{array}{ccc} [0,2^{10}[& \longrightarrow & [0,2^{10}[\\ x & \longmapsto & g(x) \end{array}$$

where g(x) is the real number of $[0,2^{10}[$ defined below:

- its integral part has a binary decomposition equal to $e'_0,\ldots,e'_9$, with:

$$e'_i = \begin{cases} e(x)_i & \text{if } i \neq s_0 \\ e(x)_i + 1 \ (\text{mod } 2) & \text{if } i = s_0 \end{cases}$$

- whose decimal part is $s(x)_1, s(x)_2, \ldots$

In other words, if $x = \sum_{k=0}^{9} 2^{9-k} e_k + \sum_{k=0}^{+\infty} s^k\, 10^{-k-1}$, then:

$$g(x) = \sum_{k=0}^{9} 2^{9-k}(e_k + \delta(k,s_0)) \ (\text{mod } 2) + \sum_{k=0}^{+\infty} s^{k+1} 10^{-k-1}.$$

Defining a metric on $[0,2^{10}[$

Numerous metrics can be defined on the set $[0,2^{10}[$, the most usual one being the Euclidian distance $\Delta(x,y) = |y-x|^2$. This Euclidian distance does not reproduce exactly the notion of proximity induced by our first distance d on $\mathcal{X}$. Indeed d is finer than Δ. This is the reason we have to introduce the following metric:

Definition 63 Let $x,y \in [0,2^{10}[$. D denotes the function from $[0,2^{10}[^2$ to $\mathbb{R}^+$ defined by: $D(x,y) = D_e\left(e(x),e(y)\right) + D_s\left(s(x),s(y)\right)$, where:

$$D_e(E,\check{E}) = \sum_{k=0}^{9} \delta(E_k,\check{E}_k), \quad \text{and} \quad D_s(S,\check{S}) = \sum_{k=1}^{\infty} \frac{|s_k - \check{S}^k|}{10^k}.$$

Proposition 28 D is a distance on $[0,2^{10}[$.

PROOF The three axioms defining a distance must be checked.

- $D \geqslant 0$, because everything is positive in its definition. If $D(x, y) = 0$, then $D_e(x, y) = 0$, so the integral parts of x and y are equal (they have the same binary decomposition). Additionally, $D_s(x, y) = 0$, then $\forall k \in \mathbb{N}^*, s(x)^k = s(y)^k$. In other words, x and y have the same k-th decimal digit, $\forall k \in \mathbb{N}^*$. And so $x = y$.

- $D(x, y) = D(y, x)$.

- Finally, the triangular inequality is obtained, due to the fact that both δ and $|x - y|$ satisfy it.

The convergence of sequences according to D is not the same as the usual convergence related to the Euclidian metric. For instance, if $x^n \to x$ according to D, then necessarily the integral part of each x^n is equal to the integral part of x (at least after a given threshold), and the decimal part of x^n corresponds to the one of x "as far as required." To illustrate this fact, a comparison between D and the Euclidian distance is given in Figure 5.1. These illustrations show that D is richer and more refined than the Euclidian distance, and thus is more precise.

The semiconjugacy

It is now possible to define a topological semiconjugacy between $\mathcal{X}$ and an interval of $\mathbb{R}$:

Theorem 15 *Chaotic iterations on the phase space $\mathcal{X}$ are simple iterations on $\mathbb{R}$, which is illustrated by the semiconjugacy of the diagram below:*

$$
\begin{array}{ccc}
\left(\mathcal{S}_{10} \times \mathbb{B}^{10}, d \right) & \xrightarrow{\;G_{f_0}\;} & \left(\mathcal{S}_{10} \times \mathbb{B}^{10}, d \right) \\
{\scriptstyle \varphi} \downarrow & & \downarrow {\scriptstyle \varphi} \\
\left([0, 2^{10}[, D \right) & \xrightarrow[\;g\;]{} & \left([0, 2^{10}[, D \right)
\end{array}
$$

PROOF φ has been constructed in order to be continuous and onto.

In other words, $\mathcal{X}$ is approximately equal to $[0, 2^{\mathsf{N}}[$.

5.5.1.2   Chaotic iterations described as a real function

It can be remarked that the function g is a piecewise linear function: it is linear on each interval having the form $\left[\dfrac{n}{10}, \dfrac{n+1}{10} \right[$, $n \in [\![0; 2^{10} \times 10]\!]$ and its slope is equal to 10. Let us justify these statements:

Proposition 29 *Chaotic iterations g defined on $\mathbb{R}$ have derivatives of all orders on $[0, 2^{10}[$, except on the 10241 points in I defined by* $\left\{ \dfrac{n}{10} \;/\; n \in [\![0; 2^{10} \times 10]\!] \right\}.$

Furthermore, on each interval of the form $\left[\dfrac{n}{10}, \dfrac{n+1}{10}\right[$, *with* $n \in [0; 2^{10} \times 10]$, g *is a linear function, having a slope equal to 10:* $\forall x \notin I, g'(x) = 10$.

PROOF Let $I_n = \left[\dfrac{n}{10}, \dfrac{n+1}{10}\right[$, with $n \in [0; 2^{10} \times 10]$. All the points of I_n have the same integral part e and the same decimal part s_0: on the set I_n, functions $e(x)$ and $x \mapsto s(x)^0$ of Definition 61 only depend on n. So all the images $g(x)$ of these points x:

- Have the same integral part, which is e, except probably the bit number s_0. In other words, this integer has approximately the same binary decomposition as e, the sole exception being the digit s_0 (this number is then either $e + 2^{10-s_0}$ or $e - 2^{10-s_0}$, depending on the parity of s_0, *i.e.*, it is equal to $e + (-1)^{s_0} \times 2^{10-s_0}$).

- A shift to the left has been applied to the decimal part y, losing by doing so the common first digit s_0. In other words, y has been mapped into $10 \times y - s_0$.

To sum up, the action of g on the points of I is as follows: first, make a multiplication by 10, and second, add the same constant to each term, which is $\dfrac{1}{10}\left(e + (-1)^{s_0} \times 2^{10-s_0}\right) - s_0$.

Remark 14 Finally, chaotic iterations used in this chapter are elements of the large family of functions that are both chaotic and piecewise linear (like the tent map [138]).

We are now able to evaluate the Lyapunov exponent of the chaotic iterations, which are now described by the iterations on $\mathbb{R}$ of the function g introduced in Definition 62.

5.5.2 Evaluation of the Lyapunov Exponent

Let $\mathcal{L} = \left\{x^0 \in [0, 2^{10}[\; / \; \forall n \in \mathbb{N}, x^n \notin I\right\}$, where I is the set of points in the real interval where g is not differentiable (as it is explained in Proposition 29). Then,

Theorem 16 $\forall x^0 \in \mathcal{L}$, *the Lyapunov exponent of chaotic iterations having* x^0 *for initial condition is equal to* $\lambda(x^0) = \ln(10)$.

PROOF It is reminded that g is piecewise linear, with a slope of 10 ($g'(x) = 10$ where the function g is differentiable). Then $\forall x \in \mathcal{L}$,

$$\lambda(x) = \lim_{n \to +\infty} \frac{1}{n} \sum_{i=1}^{n} \ln \left| g'\left(x^{i-1}\right)\right| = \lim_{n \to +\infty} \frac{1}{n} \sum_{i=1}^{n} \ln |10| =$$

$$\lim_{n \to +\infty} \frac{1}{n} n \ln |10| = \ln 10.$$

The set of initial conditions for which this exponent is not calculable, is finally countable. This is indeed the initial condition such that an iteration value will be a number having the form $\dfrac{n}{10}$, with $n \in \mathbb{N}$. We can reach such a real number only by starting iterations on a *decimal number*, as the latter must have a finite fractional part.

Remark 15 For a system having N cells, we will find, *mutatis mutandis*, an infinite uncountable set of initial conditions $x^0 \in \left[0; 2^{\mathsf{N}}\right[$ such that $\lambda(x^0) = \ln(\mathsf{N})$.

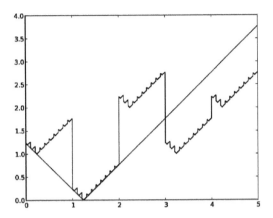

(a) Function $x \to dist(x; 1, 234)$ on the interval $(0; 5)$.

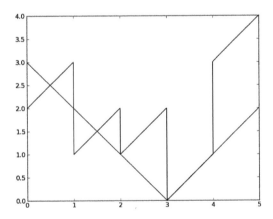

(b) Function $x \to dist(x; 3)$ on the interval $(0; 5)$.

FIGURE 5.1: Comparison between D and the Euclidian distance.

Chapter 6

Theoretical Proofs of Chaotic Machines

6.1   Chaotic Turing Machines

Let us consider a given algorithm. Because it must be computed one day, it is always possible to translate it as a Turing machine, and this last machine can be written as $x^{n+1} = f(x^n)$ in the following way. Let (w, i, q) be the current configuration of the Turing machine (Figure 6.1), where $w = \sharp^{-\omega} w(0) \ldots w(k) \sharp^{\omega}$ is the paper tape, i is the position of the tape head, q is used for the state of the machine, and δ is its transition function (the notations used here are well-known and widely used). We define f by:

- $f(w(0) \ldots w(k), i, q) = (w(0) \ldots w(i-1) a w(i+1) w(k), i+1, q')$, if $\delta(q, w(i)) = (q', a, \to)$,

- $f(w(0) \ldots w(k), i, q) = (w(0) \ldots w(i-1) a w(i+1) w(k), i-1, q')$, if $\delta(q, w(i)) = (q', a, \leftarrow)$.

Thus the Turing machine can be written as an iterate function $x^{n+1} = f(x^n)$ on a well-defined set $\mathcal{X}$, with x^0 as the initial configuration of the machine. We denote by $\mathcal{T}(S)$ the iterative process of the algorithm S.

Let τ be a topology on $\mathcal{X}$. So the behavior of this dynamical system can be studied to know whether or not the algorithm is τ-unpredictable.

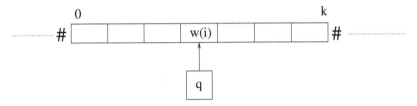

FIGURE 6.1: Turing Machine

6.2   Practical Issues

Let us now explain how it is possible to have true chaos in a finite state machine.

6.2.1   A program with a chaotic behavior

In the Section 5.2 we have proven that discrete chaotic iterations can be put in the field of discrete dynamical systems:

$$\begin{cases} x^0 \in \mathcal{X} \\ x^{n+1} = G_f(x^n), \end{cases}$$

where $(\mathcal{X}, d)$ is a metric space and G_f is a continuous function. Thus, it becomes possible to study the topological behavior of those chaotic iterations. Precisely, it has been proven that if the iterate function is based on the vectorial logical negation f_0, then chaotic iterations generate chaos according to Devaney. Therefore chaotic iterations, as Devaney's topological chaos, satisfy: sensitive dependence on the initial conditions, unpredictability, indecomposability, and uniform repartition. Additionally, G_{f_0} has been proven to be expansive and topologically mixing, and its topological entropy has been computed. Our intention is now to use these chaotic iterations, that are highly unpredictable, to build programs in the computer science security field. Furthermore, we will give in Section 7.3.4 a link between CIs and artificial neural networks, thus it is possible to make them behave chaotically.

Up to now, most of computer programs presented as chaotic lose their chaotic properties while computing in the finite set of machine numbers. The algorithms that have been presented as chaotic usually act as follows. After having received its initial state, the machine works alone with no interaction with the outside world. Its outputs only depend on the different states of the machine. The main problem which prevents speaking about chaos in this particular situation is that when a finite state machine reaches the same internal state twice, the two future evolutions are identical. Such a machine always finishes by entering into a cycle while iterating. This highly predictable behavior cannot be set as chaotic, at least as expressed by Devaney. Some attempts to define a discrete notion of chaos have been proposed, but they are not completely satisfactory and are less recognized than the notions exposed in this book.

The stated problem can be solved in the following way. The computer must generate an output O computed from its current state E *and* the current value of an input S, which changes at each iteration (Figure 6.2). Therefore, it is possible that the machine presents the same state twice, but with two future evolutions completely different, depending on the values of the input. By doing

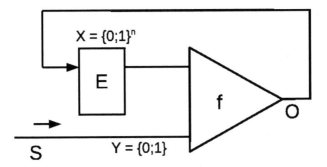

FIGURE 6.2: A chaotic finite-state machine. At each iteration, a new value is taken from the outside world (S). It is used by f as input together with the current state (E).

so, we thus obtain a machine with a finite number of states, which can evolve in infinitely different ways, due to the new values provided by the input at each iteration. Thus such a machine can behave chaotically, as defined in Devaney's formulation.

6.2.2 The practical case of finite strategies

It is worthwhile to notice that even if the set of machine numbers is finite, we deal in practice with the *infinite* set of strategies that have *finite but unbounded lengths*. Indeed, it is not necessary to store all of the terms of these strategies in the memory. Only its nth term (an integer less than or equal to N) has to be stored at the nth step, as it is illustrated in the following example. Let us suppose that a given text is input from the outside world into the computer character by character and that the current term of the strategy is computed from the ASCII code of the current stored character. Since the set of all possible texts of the outside world is infinite and the number of their characters is unbounded, we work with an infinite set of finite but unbounded strategies.

In the computer science framework, we also have to deal with a finite set of states of the form $\mathbb{B}^N$ and as stated before an infinite set $\mathbb{S}$ of strategies. The sole difference with the theoretical study is that instead of being infinite the sequences of S are finite with unbounded length, as any reasonable program must obviously finish one day.

The proofs of continuity and transitivity stated previously are independent of the finiteness of the length of strategies (sequences of $\mathbb{S}$). Sensitivity can be proven too in this situation (see Section 5.3.2.1). So even in the case of finite machine numbers, we have the two fundamental properties of chaos: sensitivity and transitivity, which respectively imply unpredictability and indecomposability (see [61], p.50). The regularity supposes that the sequences are of infinite lengths. To obtain the analogous of regularity in the context of finite sets, we can for example define a notion of *periodic but finite* sequences.

Definition 64 A strategy $S \in \mathbb{S}$ is said to be *periodic but finite* if S is a finite sequence of length n and if there exists a divisor p of n, $p \neq n$, such that $\forall i \leqslant n - p, S^i = S^{i+p}$. A point $(E, S) \in \mathcal{X}$ is said to be *periodic but finite*, if its strategy S is periodic but finite.

In this situation, $(1, 2, 1, 2, 1, 2, 1, 2)$ $(p=2)$ and $(2, 2, 2)$ $(p=1)$, are periodic but finite. This definition can be interpreted as analogous to periodicity definition on finite strategies. Following the proof of regularity (Section 5.2.2.1), it can be proven that the set of periodic but finite points is dense on $\mathcal{X}$, hence obtaining a desired element of regularity in finite sets, as quoted from Devaney ([61], p.50): "two points arbitrary close to each other could have completely different behaviors, the one could have a cyclic behavior as long as the system iterates while the trajectory of the second could 'visit' the whole phase space." It should be recalled that the regularity was introduced by Devaney in order to counteract the effects of transitivity: two points close to each other can have fundamentally different behaviors.

Part IV

Applications of chaos in the computer science field

Chapter 7

Information Security

7.1　Steganography and Digital Watermarking

7.1.1　Introduction

In past decades, the studies in the information hiding domain have almost exclusively been focused on robustness [2], [38]. Security has emerged in the last few years as a new interest in this domain [43], [106], [111]. Security and robustness are neighboring concepts without clearly established definitions [112]. Robustness is often considered to be mostly concerned with blind elementary attacks, whereas security is not limited to certain specific attacks. Security encompasses robustness and intentional attacks [56], [85]. The attempts to define the differences between robustness and security, to clarify the classes of attacks, and to give some consistency to the notion of security, illustrate the recent important concern to bring a rigorous theoretical framework for security in information hiding.

In the framework of watermarking and steganography, security has seen several important developments since the last decade [37], [51], [87]. Nevertheless, several open questions still remain. First of all, even if several security classes have been identified since the first classification of attacks, only a small part of them can be easily studied within this framework. Additionally, the current theoretical approach requires some hypotheses on the covertext, which are not always easy to check in practice, as it is detailed in the next section. In the existing approach, when a same hidden message is embedded in several covertexts, information leak is studied in terms of probability. If the leak is important, then the scheme is considered as insecure.

In this chapter we are interested in the evaluation of unpredictability of a data hiding scheme: we will consider that a scheme is secure if it is proven to be unpredictable. This approach can be used to study some classes of attacks that are difficult to investigate in the existing security approach. It also enriches the variety of qualitative and quantitative tools that evaluate how strong the security is, thus reinforcing the confidence that can be had in a given scheme. For example, let us suppose that Eve, an attacker, observes the behavior of a data hiding machine. If there is no information leak when Eve applies an input into the machine, and if the machine is unpredictable when the input changes, Eve cannot deduce anything from these observations. This vague sentence is clarified in the following sections.

7.1.2 The proposed approach compared to existing ones

7.1.2.1 Related work

The first fundamental work in security was done by Cachin in the late 90s [43] in the context of steganography. Cachin interprets the attempt of an attacker to distinguish between an innocent image and stego-content as a hypothesis testing problem. The basic properties of a stegosystem are defined using the notions of entropy, mutual information, and relative entropy. Mittelholzer [106], inspired by the work of Cachin, proposed the first theoretical framework for analyzing the security of a watermarking scheme.

These efforts to bring a theoretical framework for security in steganography and watermarking have been followed up by Kalker [85], who tries to clarify the concepts (robustness *vs.* security), and the classifications of watermarking attacks. This work has been deepened by Furon *et al.* [69], who have translated Kerckhoffs' principle (Alice and Bob shall only rely on some previously shared secret for privacy), from cryptography to data hiding. They used Diffie and Hellman methodology, and Shannon's cryptographic framework [125], to classify the watermarking attacks in categories, according to the type of information Eve has access to [51], [113]:

- Watermarked Only Attack (WOA): the attacker has access only to watermarked content.

- Known Message Attack (KMA): the attacker has access to pairs of watermarked content and corresponding hidden messages.

- Known Original Attack (KOA): occurs when an attacker has access to several pairs of watermarked content and their corresponding original versions.

- Constant-Message Attack (CMA): the attacker observes several watermarked contents and only knows that the unknown hidden message is the same in all content.

Four classes of security are defined in [50] for WOA, namely insecurity, key-security, subspace-security, and stego-security. Stego-security [50] is the highest security class in Watermark-Only Attack setup. Let $\mathbb{K}$ be the set of embedding keys, $p(X)$ the probabilistic model of N_0 initial host contents, and $p(Y|K)$ the probabilistic model of N_0 marked contents s.t. each host content has been marked with the same key K and the same embedding function.

Definition 65 (Stego-Security [50]) The embedding function is *stego-secure* if $\forall K \in \mathbb{K}, p(Y|K) = p(X)$ is established.

Finally, Cayre *et al.* have proposed in [51] the Fisher Information Matrix to quantify security in this context. However, these well-established notions cannot be extended to study the security under the KMA, KOA, and CMA setups.

7.1.2.2 Contributions of the topological approach

A novel theoretical framework for data hiding security has been proposed by the authors of this book in [68, 78], to study the KMA, KOA, and CMA setups. A data hiding scheme is considered in this proposal as a machine, whose detail is public. This machine receives: hidden messages, a secret key, and original content from the outside world and returns stego-content. In this point of view, security of the scheme depends on the unpredictable behavior of the machine: there is an indisputable lack of security when an attacker can predict where the watermark could be. To give consistency to the notion of unpredictability, this machine is modeled as a dynamical system:

$$x^0 \in \mathcal{X}, x^{n+1} = f(x^n),$$

where x^n denotes another time the nth term of the sequence x lying in a given set $\mathcal{X}$ (the connection between $x^0, \mathcal{X}, f$ and the watermarking scheme is given below). This reformulation is always possible, as it has been proven in Chapter 6, using Turing machines.

Unpredictability can thus be related to some topological or ergodic aspects of f taken from the mathematical theory of chaos, as defined by Devaney [61], Li-Yorke [92], or Adler [3], for example. This theoretical framework for some security aspects of a data hiding scheme respects Kerckhoffs' principle [88]. It is based on a topological description of data hiding, whereas most studies in this field usually have used the theory of probability [70, 112]. The goal here is to offer an additional contribution to the variety of security evaluations, which should lead to better confidence in data hiding schemes, and to solve the lack of a security notion in CMA, KOA, and KMA setups.

Compared to the information-theoretic model for steganography proposed by Cachin [43] and extended for example by Ker [87], the topological security, which will be defined in Section 7.1.3.2, can be used in KOA, KMA, and CMA setups [68, 78]. Moreover, it can check whether the claim of a chaotic behavior for a topologically based data hiding scheme can be verified or not: it is natural to check this claim and to evaluate the strength of this chaotic behavior when starting a security study of a topologically based data hiding scheme [68, 78]. Then, if all of the requirements concerning the topological security can be well-established, a stego-security study should be executed in a second stage (if possible), to validate completely the data hiding scheme.

Stego-security studies take place in the WOA category and are related to the Simmons' prisoners' problem [128]. In this problem, Alice and Bob are in jail and want to plan an escape by exchanging hidden messages through innocent-looking cover contents. These messages are conveyed by Eve, a warden, who tries to benefit from any information leak resulting from the use of the same secret key. Quoting Cayre and Bas in [50]: "Like other works, we consider Alice and Bob use only one key. Of course, in real applications, especially in steganography, it is highly desirable to change the key at every communication between Alice and Bob." In addition, a probabilistic model of the covertext is needed and, as stated by Cachin in [43], "assuming the existence of a covertext distribution seems to render our model somewhat unrealistic for the practical purpose of steganography": there is no "generic host model" that can be used in every steganalysis scenario. Even though this assumption does not rule out the probabilistic steganalysis methods (the possibly marked host is made available for a steganalist, and it can thus be classified and modeled within a determined subgroup of hosts with an accurate model and covertext distribution), it raises some technical issues which are not always easy to solve.

The framework proposed in [68,78], which is based on the theory presented in the first chapters of this book, does not suppose any assumption of this kind. It works with simple or multiple secret keys. Last, but not least, it is not restricted to the sole WOA category of attacks. This new framework can be helpful in several situations. First, it contributes to reinforce the confidence in schemes that have been proven secure with the information-theoretic tools, as it is done in Section 7.1.4. Second, this new notion of security can be used to replace Cachin's theory, when this latter cannot be used. Last, it can evaluate the claim of some proposed information hiding schemes to be chaotic. This approach is explained in detail in the following section.

7.1.3 Chaos for data hiding security

7.1.3.1 State-of-the-art

Various computer security fields, like digital watermarking, can have an interest in working in a rigorous chaos framework. This mathematical theory presented in the first chapters of this book brings some qualitative and quantitative tools, which allow the proof and evaluation of qualities required in these fields: unpredictability, disorder, uniform repartition, sensitivity, fragility, topological entropy or mixture... The security of existing algorithms can consequently be reinforced, while new chaotic methods can be discovered. However, until now, most of the methods presented lose their chaotic properties while computing in the finite set of machine numbers.

The performances and suitability of chaotic signals in digital watermarking have been proven in recent years [107]. In existing works, chaos is sometimes employed in order to encrypt the watermark, or to embed it in the carrier image. For example, binary digital signals generated by using chaotic dynamical systems are proposed as watermarks in [133], providing a sufficient watermark complexity and efficient robustness under various image processing. Other approaches can be found for example in [139], [97], or [60]. In these papers, it is implicitly understood that the use of chaotic maps as components of the scheme lead to a chaotic scheme as a whole. But this assumption is not trivial and must be proven. In addition, the definition of chaos is almost never given. This term is often confused with the sensitivity to initial conditions.

Indeed, we have shown in Part II that the notion of chaos encompasses several rigorous definitions that are not equivalent together. A scheme that is chaotic as defined by Devaney does not always behave as a chaotic scheme in the sense of the topological entropy, for instance. Each definition of chaos results from a set of chaotic properties like sensitivity, regularity, and transitivity, for Devaney. But the question "what chaotic properties are needed to achieve goals like robustness, security, or authentication?" is never elucidated in these papers. Moreover, in these works, chaos is used to have robustness, but this notion is more related to security. Authentication and encryption are not always proposed. Finally, the property of chaos is usually reduced to the simple use of a logistic map to select the embedding coefficients but, because of the finiteness of the memory of a computer, only a kind of discrete chaos is generated, and the logistic map seems to be inadequate for cryptographic applications [7]. Consequences of these approximations are never discussed.

7.1.3.2   Topological security

Let us now present in more detail the new notion of topological security. To check whether an existing data hiding scheme is chaotic or not, we propose first to write it as an iterate process $x^{n+1} = f(x^n)$ (it has been proven in Chapter 6 that such a reformulation is always possible). Then,

Definition 66 Let S be an information hiding scheme described by its iterative process $x^{n+1} = f(x^n)$, which is denoted by $\mathcal{T}(S)$. S is said to be topologically secure on $(\mathcal{X}, \tau)$ if f has a chaotic behavior, as defined by Devaney, on this topological space $(\mathcal{X}, \tau)$.

Theoretically speaking, topological security can always be studied, because it only requires that the two following points are satisfied.

- First, the data hiding scheme must be written as an iterate function on a set $\mathcal{X}$. We have stated in Chapter 6 that this is possible for any given scheme.

- Second, a metric or a topology must be defined on $\mathcal{X}$. This is always possible, for instance by taking either the discrete or the trivial topology, even though the latter is not really relevant for the results we wish to obtain.

Topological security is clearly impacted by the choice of the distance or topology on $\mathcal{X}$ and this dependence must be studied with attention. It is evident that the choice of the metric (or of the topology) must be regarded carefully, for example by establishing a strong link between the proximity of two points and the results that data hiding attempts to obtain. However, some topologies are more natural and reasonable than others, and equivalence of distances in finite dimensions reduces the impact of this choice. In addition, it can be remarked that stego-security supposes the same kind of hypothesis; dealing with probabilities implies the definition of a sigma-algebra. To our best knowledge, the Borel algebra is always chosen, even though this choice is neither stated, nor justified. As a topological space is needed to define Borel sets, we can claim that at least when a stego-security study is achieved, then a chaos study can be realized with the same topology (and the same relevance).

Topological security is the first level of our security scale based on unpredictability. This property is required, but is not sufficient: it is only the first stage of the evaluation of the unpredictable behavior of the scheme. This study must be followed by the establishment of the list of chaotic properties that the system presents. Indeed, being unpredictable is a tricky thing to define and the number of candidates that give consistency to this notion is large: topological and metric entropy, ergodicity, topological mixing, the Lyapunov exponent, expansiveness, transitivity and strong transitivity, bifurcation theory, or chaos as defined by Li-Yorke, by Devaney or by Knudsen, as they are defined in the second part of this book. As each definition illustrates a particular aspect of an unpredictable behavior and has its own interest, each notion of chaos offers new light on the security of a data hiding scheme. Thus we consider that a given data hiding scheme will be more secure than another one if it presents a larger number of chaotic qualities and if its quantitative values are better. Indeed, the chaotic properties to check depend on the data hiding objectives to reach: fragile watermarking, robustness, resistance against attacks in CMA setup, *etc.* The application of some of these qualities can be irrelevant for determined security aspects, even depending on the application to which the system is designed. This point is illustrated in the next section.

7.1.3.3 Unpredictability and classes of attacks

We consider now the four classes of attacks recalled in the beginning of this chapter. Stego-security is clearly relevant and required in a WOA setup: Eve only has access to watermarked contents and due to stego-security, it is impossible for her to decide whether a content has been processed through the embedding function or not. So in a WOA setup, a stego-secure algorithm could face Eve's attacks. However, such a framework is not as useful to counteract KOA, KMA, and CMA classes of attacks. In these setups, Eve tries to benefit from her observations of watermarked contents, when she changes some initial conditions in the data hiding scheme. She desires to have a sufficient understanding of the scheme and to be able to predict its behavior.

This knowledge can serve an attacker in various situations, for example in a man-in-the-middle attack through a hidden channel, or when trying to counteract digital rights management (DRM). Let us explain for example how Eve can try to achieve a man-in-the-middle attack by taking advantage of the predictable behavior of a scheme in the KOA setup. We suppose that Alice and Bob communicate through a hidden channel in some innocent master paintings. Eve has thus access to original and watermarked paintings, by using a base of knowledge and observing the communication channel. Let us now suppose that Alice sends a watermarked painting P_o to Bob. If Eve is able to predict the behavior of the data hiding scheme used by Alice and Bob, then she can:

1. intercept P_o,

2. use the same painting P as Alice,

3. and try to predict how her own message should change P.

4. The result of this prediction is sent to Bob.

It is true that the chances of success for this attack are low, but these chances increase with the predictability of the data hiding scheme or when the size of the hidden payload is small (a few bits), thus revealing a security failure.

Let us now suppose that Eve wants to attack some DRM. She has access to several pairs of watermarks (the copyrights, which are supposed here to be public) and watermarked media: we are in a KMA setup. In addition, she has access to the data hiding machine, but lacks knowledge of a secret key parameter (we suppose that this particular DRM scheme respects Kerckhoffs' principle). This key thus determines how to apply the copyright to the media. She wants to insert her own copyright into this protected media to make it impossible to determine whether Eve is the owner or not. She does not know exactly how copyrights are applied to the original media because, as stated before, the DRM machine is supposed to work with a secret key. However, Eve can reach her goal if she is able to predict the behavior of the copyright machine; she can approximately determine what the watermarked media should be with its own copyright. Indeed, the prediction of the behavior of the copyright machine must not be an easy task. The key should be what determines this behavior, and two given keys must lead to two completely different behaviors; this is exactly what appends with a chaotic system. Moreover, even though bypassing the security of the key must not be easy, there is a lack of security if Eve can approximately predict the behavior of the system with an approximate estimation of the key: various challenges of hackers have been achieved by obtaining such an approximation (by studying memory, *etc.*). Additionally, a copyright is not a fully random message, but contains some predictable information. Topological security can solve these issues.

In conclusion, at least in situations similar to the two examples above, unpredictable schemes are required to grant security under the proposed understanding.

7.1.4 Topological security of spread-spectrum data hiding schemes

7.1.4.1 A first proof of topological security

In what follows, the proposed framework is used to give a first topological security evaluation of the well-known spread-spectrum (SS) data hiding techniques [50]. This proves that the previous framework is ready for real-world applications and establishes a concrete link between chaos and stego-security notions. Some consequences concerning the use of SS in KOA, KMA, and CMA setups are given at the end of this section. We will now prove that,

Theorem 17 *Spread-spectrum data hiding techniques are topologically secure.*

Let $x \in \mathbb{R}^{N_v}$ be a host vector in which we want to hide a message $m \in \{0; 1\}^{N_c}$. N_c is the size of the hidden payload (in bits) and N_v the size of the stego or host vector (in samples). A key $\mathcal{K}$ is used to initialize a pseudorandom number generator to obtain N_c secret carries $\{u^i\}$ taken in $\mathbb{R}^{N_v}$, which can be supposed to be orthonormalized. Thus in classical SS the watermark signal w is constructed as follows [50]:

$$w = \sum_{i=0}^{N_c-1} \gamma(-1)^{m^i} u^i, \tag{7.1}$$

where γ is a given distortion level. The watermarked signal y is then defined by [50]:

$$y = x + w. \tag{7.2}$$

Let us now suppose that the components of the watermark are bounded by a finite value N_b, that is, $\max(\{w_i, i \in [\![1, N_v]\!]\}) \leqslant \mathsf{N}_b$. This bound can be as large as needed; however, a very large N_b seems to be contradictory to the aims of a data hiding scheme. Let us consider $\mathcal{X} = \left([-\mathsf{N}_b, \mathsf{N}_b]^{N_v} \right)^{\mathbb{N}} \times \mathbb{R}^{N_v}$ and

$$G((S, E)) = (\sigma(S); i(S) + E), \tag{7.3}$$

where σ is the *shift* function defined by $\sigma : (S^n)_{n \in \mathbb{N}} \in \left([-\mathsf{N}_b, \mathsf{N}_b]^{N_v} \right)^{\mathbb{N}} \rightarrow$ $(S^{n+1})_{n \in \mathbb{N}} \in \left([-\mathsf{N}_b, \mathsf{N}_b]^{N_v} \right)^{\mathbb{N}}$ and the *initial function* i is the map which associates to a sequence, its first term: $i : (S^n)_{n \in \mathbb{N}} \in \left([-\mathsf{N}_b, \mathsf{N}_b]^{N_v} \right)^{\mathbb{N}} \rightarrow S^0 \in$ $[-\mathsf{N}_b; \mathsf{N}_b]^{N_v}$. E will be the vector describing the part of the host that can be altered without sensitive damages, when S will give the location of the alteration at each iteration (S will depend on the hidden message and the secret key).

Spread-spectrum data hiding techniques are thus the result of N_c iterations of the following dynamical system:

$$\begin{cases} X^0 \in \mathcal{X}, \\ X^{n+1} = G(X^n), \end{cases} \tag{7.4}$$

and the watermarked media is the second component of X^{N_c}. Indeed, the second component of X^k corresponds to the host image after k alterations, whereas the first component explains how to alter it another time.

Classical SS (*i.e.*, with BPSK modulation [50]) is defined by $X^0 = (S^0, E^0)$ where E^0 is the host vector x and S^0 is the sequence

$$\left((-1)^{m^0} \gamma\, u^0, (-1)^{m^1} \gamma\, u^1, \ldots, (-1)^{m^{N_c - 1}} \gamma\, u^{N_c - 1} \right), \qquad (7.5)$$

in which γ allows to achieve a given distortion, whereas in ISS (Improved Spread Spectrum [102]), S^0 is defined by

$$\left((-1)^{m^i} \alpha - \lambda \frac{<x, u^i>}{||u^i||^2} \right)_{i=0,\ldots,N_c-1}, \qquad (7.6)$$

where α and λ are computed to achieve an average distortion and to minimize the error probability [50]. Last, in natural watermarking NW, S^0 is defined by

$$\left(-\left(1 + \eta(-1)^{m^i} \frac{<x, u^i>}{|<x, u^i>|} \right) \frac{<x, u^i>}{||u^i||^2} \right)_{i=0,\ldots,N_c-1}. \qquad (7.7)$$

This modulation consists in a model-based projection on the different vectors u^i followed by a η-scaling along the direction of u^i. Natural watermarking has been proven stego-secure when $\eta = 1$ (see [50]), so we can claim that,

Proposition 30 *Spread-spectrum G can be a stego-secure data hiding technique.*

We will now prove in what follows that spread-spectrum data hiding schemes are topologically secure, thus finding in SS a first link between chaos and stego security.

Let $d_\infty(A, B) = max\left\{ |A_i - B_i|, i = 1 \ldots N_v \right\}$ be the usual metric on $\mathbb{R}^{N_v}$. We define a new distance between two points $X = (S, E), Y = (\check{S}, \check{E}) \in \mathcal{X}$ by

$$d(X, Y) = d_\infty(E, \check{E}) + d_s(S, \check{S}), \qquad (7.8)$$

where $d_s(S, \check{S}) = \dfrac{9}{N_b} \displaystyle\sum_{k=0}^{\infty} \dfrac{d_\infty(S^k, \check{S}^k)}{10^k}$.

The choice of d_∞ on $\mathbb{R}^{N_v}$ is not important, because of the equivalence of metrics in finite dimension: topologies are the same, thus chaos properties are not changed by using another distance on $\mathbb{R}^{N_v}$. d_s has been chosen such that $d(X, Y)$ is small when the distance between the watermarked images resulting on the spread-spectrum applied on X and Y are close (for any metrics on $\mathbb{R}^{N_v}$, as they are all equivalent). Last, $\dfrac{9}{N_b}$ is just a normalization value.

We will now prove that:

Proposition 31 *G is continuous on $(\mathcal{X}, d)$.*

PROOF We use the sequential continuity. Let $(S_n, E_n)_{n \in \mathbb{N}}$ be a sequence of the phase space $\mathcal{X}$, which converges to (S, E). We will prove that $(G(S_n, E_n))_{n \in \mathbb{N}}$ converges to $G(S, E)$. Let us recall that for all n, S_n is a strategy, thus, we consider a sequence of strategies (*i.e.*, a sequence of sequences).

As $d((S_n, E_n); (S, E))$ converges to 0, each distance $d_\infty(E_n, E)$ and $d_s(S_n, S)$ converges to 0.

1. If $\dfrac{9}{N_b} \displaystyle\sum_{k=0}^{\infty} \dfrac{d_\infty(S_n^k, S^k)}{10^k} \to 0$ when $n \to \infty$, then $\dfrac{9}{N_b} \displaystyle\sum_{k=1}^{\infty} \dfrac{d_\infty(S_n^k, S^k)}{10^k} \to 0.$

   So $\dfrac{9}{N_b} \displaystyle\sum_{k=0}^{\infty} \dfrac{d_\infty(S_n^{k+1}, S^{k+1})}{10^{k+1}} = \dfrac{1}{10} d_s(\sigma(S_n); \sigma(S)) \to 0.$ As a consequence, $d_s(\sigma(S_n), \sigma(S))$ converges to 0.

2. Let us prove that $d_\infty\left(i(S_n) + E_n; i(S) + E\right) \to 0.$

$$d_\infty\left(i(S_n) + E_n; i(S) + E\right)$$

$$= max\left\{\left|(i(S_n)_k + (E_n)_k) - (i(S)_k + E_k)\right|, k = 1 \ldots N_v\right\}$$

$$= max\left\{\left|(i(S_n)_k - i(S)_k) + ((E_n)_k - E_k)\right|, k = 1 \ldots N_v\right\}$$

$$\leqslant max\left\{\left|i(S_n)_k - i(S)_k\right| + \left|(E_n)_k - E_k\right|, k = 1 \ldots N_v\right\}$$

$$\leqslant max\left\{\left|i(S_n)_k - i(S)_k\right|, k = 1 \ldots N_v\right\} + d_\infty(E_n, E)$$

$$= d_\infty(S_n^0, S^0) + d_\infty(E_n, E)$$

$$\leqslant d_s(S_n, S) + d_\infty(E_n, E)$$

$$= d\left((S_n, E_n); (S, E)\right) \to 0.$$

Proposition 32 *Periodic points of G are dense in $(\mathcal{X}, d)$, so G is regular.*

PROOF Let $(S, E) \in \mathcal{X}$ and $\varepsilon > 0$. We are looking for a periodic point $(\check{S}, \check{E}) \in \mathcal{X}$ such that $d\left((S, E), (\check{S}, \check{E})\right) < \varepsilon$. Let $\check{E} = E$ and S_n denotes the sequence defined by:

$$\begin{cases} S_n^k = S^k & \forall k \leqslant n \\ S_n^k = (N_b, \ldots, N_b) & \text{if } k > n \text{ and } k \equiv 0 \ (\text{mod } 2) \\ S_n^k = (-N_b, \ldots, -N_b) & \text{else.} \end{cases}$$

Then $d_s(S_n, S) = \dfrac{9}{N_b} \displaystyle\sum_{k=n+1}^{\infty} \dfrac{d_\infty(S_n^k, S^k)}{10^k} \leqslant \dfrac{9}{N_b} \displaystyle\sum_{k=n+1}^{\infty} \dfrac{N_b}{10^k} = \dfrac{1}{10^n} \to 0$ when $n \to \infty$. So $\exists n_0 \in \mathbb{N}$ such that $d_s(S^{n_0}, S) < \varepsilon$. The point (S^{n_0}, E) is then a periodic point of $\mathcal{X}$ which is ε-close to the given point (S, E).

We will now prove that,

Proposition 33 *G is transitive on $(\mathcal{X}, d)$.*

PROOF Let $B_A = \mathcal{B}(X_A, r_A)$ and $B_B = \mathcal{B}(X_B, r_B)$ be two open balls of $\mathcal{X}$, where $X_A = (S_A, E_A)$ and $X_B = (S_B, E_B)$. We are looking for $\tilde{X} = (\tilde{S}, \tilde{E}) \in B_A$ such that $\exists n_0 \in \mathbb{N}, G^{n_0}(\tilde{X}) \in B_B$.

Let $k_0 \in \mathbb{Z}$ such that $10^{-k_0} \leqslant r_A < 10^{-k_0+1}$ and $\left(\check{S}, \check{E}\right) = G^{k_0}(X_A)$. We define $\tilde{X} = (\tilde{S}, \tilde{E})$ as below:

- $\tilde{E} = E_A$,

- $\forall k \leqslant k_0, \tilde{S}^k = S_A^k$,

- $\forall k \in [\![1, N_v]\!], \tilde{S}^{k_0+k} = (-\check{E}^k + E_B^k) \times (0, \ldots, 0, 1, 0, \ldots, 0)$, *i.e.* the vector $\tilde{S}^{k_0+k}$ has its components null, except its kth, equals to $(-\check{E}^k + E_B^k)$,

- $\forall k \in \mathbb{N}, \tilde{S}^{k_0+N_v+k+1} = S_B^k$.

With such a definition, $\tilde{X}$ is in B_A and satisfies $G^{k_0+N_v}\left(\tilde{X}\right) \in B_B$.

As G is regular and transitive on $(\mathcal{X}, d)$, we can conclude that G is chaotic as defined by Devaney, thus proving Theorem 17.

7.1.4.2   Qualitative and quantitative evaluation of spread-spectrum

As stated before, the proof that a given data hiding scheme is topologically secure is just the beginning of the study. It does not allow to determine in which class (KOA, KMA, CMA, or WOA) the considered scheme can be used without security issues. The next stage is to evaluate the quality of its chaotic behavior, by using the numerous qualitative and quantitative tools offered by the theory of chaos. These tools help to compare two given topologically secure schemes, by deciding which scheme is the most unpredictable for a given class of attack, and thus must be preferred.

To illustrate, some tools are studied in this section, namely strong transitivity, and the constants of expansiveness and sensitivity. We will use them to give a better understanding of the unpredictability of spread-spectrum techniques under the KOA, KMA, and CMA setups.

Let us first prove that,

Proposition 34 *Spread-spectrum G is strongly transitive on $(\mathcal{X}, d)$.*

PROOF Let us reconsider the proof of the transitivity of G in $(\mathcal{X}, d)$. We have defined $\tilde{X} \in B_A$ such that $G^{k_0+N_v}\left(\tilde{X}\right) \in B_B$. Indeed, for this $\tilde{X}$, we have:
$$G^{k_0+N_v}\left(\tilde{X}\right) = X_B.$$

This property reinforces the effects and consequences of the transitivity, in terms of topological security and authentication. Indeed, due to strong transitivity, the set of watermarked media, obtained when using a fixed watermark, is potentially equal to the whole set of media. In that situation, Eve cannot divide the set of media to be studied, thus reducing the interest of a Constant-Message Attack.

We show now that spread spectrum techniques possess the property of topological mixing, which reinforces the strong transitivity. The consequences of this property are discussed in Section 7.1.5.3. We have the result,

Proposition 35 *Spread-spectrum techniques G are topologically mixing on $(\mathcal{X}, d)$.*

This result is an immediate consequence of the lemma below.

Lemma 2 *For any open ball B of $\mathcal{X}$, an index n can be found such that $G^n(B) = \mathcal{X}$.*

PROOF Let $B = B((E, S), \varepsilon)$ be an open ball, on which the radius can be considered as strictly less than 1. All the elements of B have the same state E and are such that an integer $k \left(= -\log_{10}(\varepsilon)\right)$ satisfies:

- all the strategies of B have the same k first terms,

- after the index k, all values are possible.

Then, after k iterations, the new state of the system is $G^k(E, S)_1$ and all the strategies are possible (all the points $(G^k(E, S)_1, \hat{S})$, with any $\hat{S} \in \mathbb{S}$, are reachable from B).

We will prove that all points of $\mathcal{X}$ are reachable from B. Let $(E', S') \in \mathcal{X}$. So the point $(\check{E}, \check{S})$ of B defined by:

- $\check{E} = E$,

- $\check{S}^i = S^i, \forall i \leqslant k$,

- $\check{S}^{k+i} = -E^i + E'^i, \forall i \leqslant N_v$,

- $\forall i \in \mathbb{N}, S^{k+N_v+i} = S'^i$.

is such that $G^{k+N_v}(\check{E}, \check{S}) = (E', S')$. This concludes the proofs of the lemma and of the proposition.

We now investigate some quantitative properties of the information hiding schemes under consideration. The constant of sensitivity δ can be obtained by proving the sensitivity without Banks' theorem.

Proposition 36 *Spread-spectrum data hiding techniques have sensitive dependence on initial conditions on $(\mathcal{X}, d)$ and the constant of sensitivity is at least equal to $\dfrac{\mathsf{N}_b}{2}$.*

PROOF Let $X = (S, E) \in \mathcal{X}$, $B = \mathcal{B}(X, r)$ an open ball centered in X, and $k_0 \in \mathbb{Z}$ such that $10^{-k_0} \leqslant r < 10^{-k_0+1}$. We define $\check{X}$ by:

- $\check{E} = E$,

- $\check{S}^k = S^k$, $\forall k \in \mathbb{N}$ such that $k \neq k_0 + 1$,

- if $S_1^{k_0+1} < \frac{N_b}{2}$, then $\check{S}_1^{k_0+1} = N_b$, else $\check{S}_1^{k_0+1} = 0$,

- $\forall i \in [\![2, N_v]\!]$, $\check{S}_i^{k_0+1} = S_i^{k_0+1}$.

So $d(X, \check{X}) = D_\infty(E, \check{E}) + d_S(S, \check{S}) = 0 + \dfrac{9}{N_b} \dfrac{d_\infty(S^{k_0+1}, \check{S}^{k_0+1})}{10^{k_0+1}} \leqslant \dfrac{9}{N_b} \dfrac{N_b}{10^{k_0+1}} \leqslant \dfrac{1}{10^{k_0}} \leqslant r$, then $\check{X} \in B$. Let us now define $\mathcal{E} : \mathcal{X} \to \mathcal{X}$, $(S, E) \mapsto E$. So $\mathcal{E}\left(G^{k_0+1}(X)\right)_0 = \mathcal{E}\left(G^{k_0+1}(\check{X})\right)_0$, because $E = \check{E}$ and $S^k = \check{S}^k$, $\forall k \leqslant k_0 + 1$. As:

- $\mathcal{E}\left(G^{k_0+2}(X)\right)_0 = \mathcal{E}\left(G^{k_0+1}(X)\right)_0 + S_0^{k_0+1}$,

- $\mathcal{E}\left(G^{k_0+2}(\check{X})\right)_0 = \mathcal{E}\left(G^{k_0+1}(\check{X})\right)_0 + \check{S}_0^{k_0+1}$,

- $\left| S_0^{k_0+1} - \check{S}_0^{k_0+1} \right| \geqslant \dfrac{N_b}{2}$.

We thus have

$$
\begin{aligned}
d\left(G^{k_0+2}(X), G^{k_0+2}(\check{X})\right) &\geqslant d_\infty\left(\mathcal{E}\left(G^{k_0+2}(X)\right), \mathcal{E}\left(G^{k_0+2}(\check{X})\right)\right) \\
&\geqslant \left| \mathcal{E}\left(G^{k_0+2}(X)\right)_0 - \mathcal{E}\left(G^{k_0+2}(\check{X})\right)_0 \right| \\
&\geqslant \frac{N_b}{2}.
\end{aligned}
$$

Let us now prove that,

Proposition 37 *Spread-spectrum G is not an expansive chaotic system on $(\mathcal{X}, d)$.*

PROOF Let $\varepsilon > 0$. We define:

- $X = \left(O_{N_v}; (O_{N_v}, O_{N_v}, \ldots, O_{N_v}, \ldots)\right)$,

- $Y = \left(O_{N_v}; (\frac{\varepsilon}{2} I_{N_v}, -\frac{\varepsilon}{2} I_{N_v}, \ldots, \frac{(-1)^n \varepsilon}{2} I_{N_v}, \ldots)\right)$,

where $O_{N_v} = (0, \ldots, 0)$ is the null vector of size N_v and I_{N_v} is the vector of size N_v equal to $(1, 0, \ldots, 0)$. Thus, for these two points, we have: $\forall n \in \mathbb{N}, d\left(G^n(X); G^n(Y)\right) < \varepsilon$.

TABLE 7.1: Evaluation of spread-spectrum schemes

	Property	Spread spectrum
CAPABILITY	Blind watermarking	Yes
	Domain	Spatial, frequency
	Authentication	?
	Watermark encryption	?
	Robustness	No
	topological security	Spread-spectrum
	Devaney's chaos	Yes
	Phase space	$\mathcal{X} = \left([\![-\mathsf{N}_b ; \mathsf{N}_b]\!]^{N_v} \right)^{\mathbb{N}} \times \mathbb{R}^{N_v}$
	Topology	d
	Qualitative Properties	Spread-spectrum
topological SECURITY	Strong transitivity	Yes
	Topological mixing	Yes
	Quantitative Properties	Spread-spectrum
	Sensitivity	Yes. $\delta \geqslant \dfrac{\mathsf{N}_b}{2}$
	expansiveness	No
	Property	Spread-spectrum
STEGO-SECURITY	Stego-security (SS)	Natural watermarking is SS when $\eta = 1$

7.1.4.3   Consequences

We have summed up some security aspects of spread-spectrum data hiding schemes in Table 7.1, and we now intend to discuss various consequences of the previous study.

All the variety of SS techniques are concerned with the property of topological security. In the point of view presented above, the choice of NW instead of ISS only affects the initial condition of the iterations of G. Indeed, the theory of chaos gives a global approach of the unpredictable behavior of a given system, but does not explain how to choose a "good initial condition." For example, the logistic map presented previously has a chaotic behavior, but if we choose $X^0 = 0$, then $\forall n \in \mathbb{N}, X^n = 0$: the subset of initial values must be regarded carefully. Stego-security is helpful to determine good initial conditions in WOA setup. For example S^0 defined as in Equation 7.7 with $\eta = 1$ leads to a chaos and stego secure spread-spectrum scheme (this is the Natural Watermarking scheme). To sum up, only the natural watermarking with $\eta = 1$ can be used in the WOA setup, because it is the sole class of spread-spectrum that is stego-secure. On the contrary, as spread-spectra are topologically secure, we cannot reject their use in the CMA setup.

We will now discuss some consequences of the fact that spread-spectrum is not expansive. The property of expansiveness reinforces drastically the sensitivity in the aims of reducing the benefits that Eve can obtain from an attack in KMA or KOA setups. For example, it is impossible to have an estimation of the watermark by moving the message (or the cover) as a cursor in a situation of expansiveness; the cursor will be much too sensitive and the changes will be much too important to extract any knowledge of it. On the contrary, a very large constant of expansiveness ε is unsuitable; the cover media will be strongly altered, whereas the watermark would be undetectable. Indeed, let us consider the same cover E twice with two different watermarks S, S'. Thus $d(X, Y) < 1$, where $X = (E, S), Y = (E, S')$ and d is for the distance previously defined. However, due to expansiveness, $\exists n \in \mathbb{N}, d\left(G^n(X); G^n(Y)\right) \geqslant \varepsilon$. Thus, $d_\infty \left(G^n(X)_1; G^n(Y)_1\right) \geqslant \varepsilon - 1$, so either $d_\infty \left(X_1; G^n(X)_1\right) \geqslant \dfrac{\varepsilon - 1}{2}$, or $d_\infty \left(Y_1; G^n(Y)_1\right) \geqslant \dfrac{\varepsilon - 1}{2}$. If ε is large, then at least one of the two watermarked media will be very different than its original cover. As a conclusion, a topologically secure data hiding scheme that is not expansive can possibly only be used in WOA and CMA setups, whereas KMA and KOA setups require expansiveness to be secure.

Finally, spread-spectrum (natural watermarking, $\eta = 1$) is relevant when a discrete and secure data hiding technique is required in a WOA setup. The particular case of the CMA setup must be regarded more carefully, depending on the situation under consideration. Last, these techniques should not be used in KOA and KMA setups, due to their lack of expansiveness.

In the next section, we will present an original class of data hiding schemes [17, 68], which are expansive.

7.1.5 A class of expansive data hiding schemes based on chaotic iterations

Our goal in this section is to construct a data hiding scheme, which will be better than spread-spectrum techniques in KOA and KMA setups. To do so, we have searched in [17, 68] for a chaotic iterative process for data hiding that will be expansive. This section presents the results of our questioning, organized as follows. The data hiding algorithm based on chaotic iterations is detailed in Subsection 7.1.5.1 and two illustrative examples are given in Subsection 7.1.5.2, namely in the spatial and frequency domains. The next subsection explains why the chaotic behavior of the proposed algorithm is preserved while computing. Finally, in Subsection 7.1.5.3 the qualitative and quantitative chaotic properties of this new class of schemes are studied.

7.1.5.1 A topologically secure data hiding algorithm

The algorithm presented in this section, based on chaotic iterations, has been formerly introduced in [18]. A first approach of its robustness is given in [18] and [76], whereas its stego-security is proven in [78]. In this chapter, we extend this work and we give two original illustrative examples.

In the following, stages defining the topologically secure data hiding scheme are given in detail. To do so, we must first introduce the notions of most and least significant coefficients.

The most and least significant coefficients

Let us define the notions of most and least significant coefficients of an image (for a more formal and rigorous approach of the proposed chaos-based information hiding scheme, the reader is referred to [8, 75]).

Definition 67 For a given image, the most significant coefficients (in short, MSCs), are coefficients that allow the description of the relevant part of the image, *i.e.* its richest (in terms of embedding information), through a sequence of bits.

For example, in a spatial description of a grayscale image, a definition of MSCs can be the sequence constituted by the first four bits of each pixel (Figure 7.1). In a discrete cosine frequency domain description, each 8×8 block of the carrier image is mapped to a list of 64 coefficients. The energy of the image is mostly contained in a determined part of them, which can constitute a possible sequence of MSCs.

Definition 68 By least significant coefficients (LSCs), we mean a translation of some insignificant parts of a medium in a sequence of bits (insignificant can be understood as: "which can be altered without sensitive damages").

These LSCs can be, for example, the last three bits of the gray level of each pixel (Figure 7.1). Discrete cosine, Fourier, and wavelet transforms can also be used to generate LSCs and MSCs. Moreover, these definitions can be extended to other types of media. Finally, let us remark that the set of MSCs is not necessarily the complement of the set of LSCs, *i.e.*, the set UC of unused coefficients which are neither MSCs, nor LSCs, is not always an empty set.

  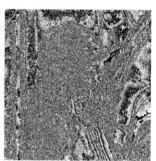

| (a) Original image (Lena). | (b) MSCs: first four bits of each pixel. | (c) LSCs: last three bits (factor 17). |

FIGURE 7.1: Example of most and least significant coefficients of Lena.

Let us now introduce the notion of authentication defined in this context:

Definition 69   [?] [?] Let $S = (\mathbb{R}^c, \mathbb{R}^w, (\mathcal{M}, \mathcal{L}, \mathcal{U}), \mathcal{A}, \mathcal{K})$ be an information hiding scheme, where:

- $\mathcal{K}$ is a set of secret keys,

- $\mathbb{R}^c$ is the set of the cover media,

- $\mathbb{R}^w$ is the set of the watermarks,

- $(\mathcal{M}, \mathcal{L}, \mathcal{U})$ are the sets of MSCs, LSCs, and UCs respectively (possibly empty), *i.e.*, $\mathcal{M} = \mathbb{R}^m$, $\mathcal{L} = \mathbb{R}^l$, and $\mathcal{U} = \mathbb{R}^{c-m-l}$. In particular, $\mathbb{R}^c = \mathcal{M} \times \mathcal{L} \times \mathcal{U}$.

- The embedding function is $e : \mathbb{R}^c \times \mathbb{R}^w \times \mathcal{K} \to \mathbb{R}^c, ((M, L, U), W, K) \mapsto (M, \mathcal{A}(K, M, L, W), U)$, where $\mathcal{A}$ is the embedding algorithm. That is, the LSCs L of the cover media (M, L, U) are set to $\mathcal{A}(K, M, L, W)$.

In this situation, S is said to be an "unauthenticated" watermarking algorithm when

$$\forall (K, M, L, W) \in \mathcal{K} \times \mathcal{M} \times \mathcal{L} \times \mathbb{R}^w, \mathcal{A}(K, M, L, W) = \mathcal{A}(K, L, W),$$

i.e., when the embedding algorithm does not depend on the MSCs. On the contrary, we said that S is an "authenticated" watermarking scheme.

A reasonable authentication is reached when at least one of the two following situations occurs:

1. When $\forall(K, L, W) \in \mathcal{K} \times \mathcal{L} \times \mathbb{R}^w, M \mapsto f(K, M, L, W)$ is an injective function.

2. When $f(K, M, L, W)$ is largely modified when a slight alteration of M is done. For example, in an expansiveness scenario, or at least when f is highly sensitive to its initial condition. In this chapter, we will focus on this situation.

In the following paragraphs, LSCs are used during the embedding stage. Indeed, some of the least significant coefficients of the carrier image will be chaotically chosen and either switched, or replaced, by the bits of the watermark. The switch should be chosen in the context of a watermark detection problem, whereas the replacement is probably more useful in pure steganography. The MSCs are only used in case of authentication; mixture and embedding stages depend on them. Hence, a coefficient should not be defined at the same time as a MSC and a LSC; the latter can be altered while the first is needed to extract the watermark.

LSCs and MSCs can appear as the converse approach of Cox & Miller feature vector extraction [58]. However, as stated before, we do not suppose that the set of MSCs is the complement of the set of LSCs. Moreover, LSCs and MSCs can be secretly shared before embedding (by using either a secret channel, or a public cryptosystem), so a pirate cannot focus only on the MSCs.

Stages of the algorithm

Our new class of data hiding schemes, called *dhCI* (data hiding scheme based on chaotic iterations) consists of two stages: (1) mixture of the watermark and (2) its embedding.

Watermark mixture. First, if required, the watermark can be mixed before its embedding in the image (Figure 7.3). A common way to achieve this stage is to use the bitwise exclusive or (XOR), for example between the watermark and a given binary sequence which behaves randomly, like a logistic map. However, as stated before, the use of this map for encryption has recently revealed several issues, and should be discouraged. In this section, we introduce a new mixture scheme based on chaotic iterations. Its chaotic strategy will be highly sensitive to the MSCs, in the case of an authenticated watermark. For the detail of this stage see Section 7.1.5.2 below.

Watermark embedding. Some LSCs will be switched, or substituted by the bits of the possibly mixed watermark (Figure 7.3). To choose the sequence of LSCs to be altered, a number of integers, less than or equal to the number N of LSCs corresponding to a chaotic sequence $\left(U^k\right)_k$, is generated from the chaotic strategy used in the mixture stage. Thus, the U^k-th least significant coefficient of the carrier image is either switched, or substituted by the kth bit of the possibly mixed watermark. In case of authentication, such a procedure leads to a choice of the LSCs which are highly dependent on the MSCs.

On the one hand, when the switch is chosen, the watermarked image is obtained from the original image whose LSBs $L = \mathbb{B}^N$ are replaced by the result of some chaotic iterations. Here, the iterate function is the vectorial

Boolean negation, the initial state is L, and strategy is equal to $\left(U^k\right)_k$. In this case, the embedding stage is reduced to chaotic iterations, so this stage is topologically secure. However, the original medium is needed to extract the watermark. This non-blind watermarking scheme can be useful to link network flows in intrusion detection as well as anonymity [82]. Another example of the use of a non-blind watermarking is given in [63], in order to protect the content of digital data through the Internet.

On the other hand, when the selected LSCs are substituted by the watermark, its extraction can be done without the original cover (blind steganography). In this case, the selection of LSBs still remains chaotic because of the use of a chaotic map, but the whole process does not satisfy topological chaos. The use of chaotic iterations is reduced to the mixture of the watermark and the embedding stage cannot be claimed topologically secure. See Section 7.1.5.2 for more detail.

Extraction. The chaotic strategy can be regenerated even in the case of an authenticated watermarking because the MSCs have not been changed during the stage of the watermark embedding. Thus, the few altered LSCs can be found, the mixed watermark can be rebuilt, and the original watermark can be obtained. In case of a switch, the result of the previous chaotic iterations on the watermarked image should be the original cover. The probability of being watermarked decreases when the number of differences increases. A ROC curve can be used to determine whether the image is watermarked or not.

In case of authentication, if the watermarked image is attacked, then the MSCs will change. Consequently, due to the high sensitivity of the embedding sequence, the LSCs designed to receive the watermark will be completely different. Hence, the result of the recovery will have no similarity with the original watermark.

In the following subsections, two application examples of the above topologically secure data hiding method are given, namely in the spatial and frequency domains. The aim is not to propose definitive or ready-made real-world applications. These case studies just enable us to explain how to encrypt the secret information using chaotic iterations, how to choose the embedding coefficients, and then how to modify these coefficients using both chaotic iterations and the encrypted message. By doing so, a first application of the theory proposed in this book is given with some illustrations.

7.1.5.2   Illustrative examples

Spatial watermarking

Images description. The carrier image is the famous Lena, which is here a 256 grayscale image of size 256×256 (Figure 7.1a). The watermark is the 64×64 pixels binary image depicted in Figure 7.2a. Notice that there is no justification for the watermark to be a binary image; the data hiding scheme works with any type of digital watermark.

The embedding domain will be the spatial domain. The selected MSCs are the four most significant bits of each pixel and the LSCs are the three last bits (a given pixel will at most be modified by four levels of gray per iteration, see Figure 7.1). Before its embedding, the watermark is encrypted by chaotic iteration. The system to iterate, chaotic strategy S^n, and the iterate function are defined below.

(a) Watermark   (b) Watermarked Lena   (c) Differences with original

FIGURE 7.2: Watermark and watermarked Lena.

Encryption of the watermark. Let us explain how to encrypt the watermark by using chaotic iterations defined in Chapter 5. The initial state x^0 of the system is constituted by the watermark considered as a Boolean vector. The iteration function is the vectorial logical negation f_0 and the chaotic strategy $(S^k)_{k \in \mathbb{N}}$ will depend on whether an authenticated watermarking method is desired or not, as follows.

Let $p \in]0, 0.5[$ be a control parameter and F denote the piecewise linear chaotic map [7], [126]:

$$F(x, p) = \begin{cases} x/p, & x \in [0, p), \\ (x - p)/(1/2 - p) & x \in [p, 1/2], \\ F(1 - x, p) & x \in [1/2, 1]. \end{cases} \quad (7.9)$$

A chaotic Boolean vector $(B^k)_{k \leqslant T}$ is generated by T iterations of F, as follows: if $x^k > \mathcal{Y}$, then $B^k = 1$, else $B^k = 0$, where $\mathcal{Y} \in]0, 1[$ is a threshold, $x^0 \in [0; 1]$, and $x^{n+1} = F(x^n, p)$. Another way to define B^k can be to use the chaotic pseudo-random number generator studied in [28] and [136], which is cryptographically secure.

Then, in case of unauthenticated watermarking, the bits of the chaotic Boolean vector (B^k) are grouped twelve by twelve, to obtain a sequence $(S^k)_{k \in \mathbb{N}}$ of integers lower than 4096, which will constitute the chaotic strategy. In case of authentication, the bitwise exclusive or (XOR) is made between the chaotic Boolean vector and the MSCs. The result is converted into a chaotic strategy by joining its bits as above. Thus, the encrypted watermark is the last Boolean vector generated by the chaotic iterations.

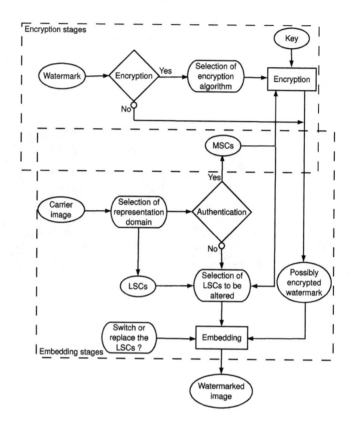

FIGURE 7.3: The topologically secure data hiding algorithm.

As presented above, the keys of this encryption scheme depend on the following data: p, $\mathcal{Y}$, X_0 and T. If we suppose that the machine numbers are coded with N bits, then a first approximation of the size of the key set is $2^{4 \times N}$.

However, this encryption scheme can be given in a more general formulation. Indeed, this cryptosystem is constituted by chaotic iterations on $\mathcal{X}' = [1, \mathsf{N}]^{\mathbb{N}} \times \mathbb{B}^{\mathsf{N}}$. In this situation, a private key is constituted by the definition of a sequence of integers $\leqslant \mathsf{N}$, a Boolean vector of size N, and a number T of iterations. So with the same assumption as above, the size of the key set is equal to $2^{\mathsf{N}} \times N$, multiplied by the cardinality of the set of sequences which have a Kolmogorov complexity lesser than a given frontier. Finally the choice of the iterate function can be integrated into the key set, if needed.

Embedding of the watermark. To embed the watermark, the sequence $(U^k)_{k \in \mathbb{N}}$ of altered bits taken from the LSCs must be defined. To do so, the strategy $(S^k)_{k\in\mathbb{N}}$ of the encryption stage is used as follows:

$$\begin{cases} U^0 &= S^0, \\ U^{n+1} &= S^{n+1} + 2 \times U^n + n \ (\text{mod } \mathsf{N}), \end{cases} \tag{7.10}$$

to obtain the result depicted in Figure 7.2b. In this example, LSBs selected by U^k are replaced by those of the watermark. The map $\theta \mapsto 2\theta$ of the torus, which is a famous example of topological Devaney's chaos [61], has been chosen to make $(U^k)_{k\in\mathbb{N}}$ highly sensitive to the chaotic strategy. As a consequence, $(U^k)_{k\in\mathbb{N}}$ is highly sensitive to the alteration of the MSCs: in case of authentication, any significant modification of the watermarked image will lead to a completely different extracted watermark.

Wavelets topologically secure watermarking

In this section, the use of our algorithm in frequency DWT domain is detailed. The same stages as in Section 7.1.5.2 will be followed to reach an identical goal. The major difference will be the way to define LSCs and MSCs.

The carrier image and watermark are the same as in the first case study, but Lena is now constituted by 512×512 pixels, to enlarge the given payload. The embedding domain is the discrete wavelets domain (DWT). In this section, the Daubechies family of wavelets is chosen: Lena is converted into its Daubechies-1 DWT coefficients, which are altered by chaotic iterations. The watermark is encrypted by chaotic iterations before its embedding. The system to iterate, chaotic strategy S^n, and iterate function are defined as before.

The algorithm depends on a decomposition level and a coefficient matrix (Figure 7.4): LL means approximation coefficient, when HH, LH, HL denote respectively diagonal, vertical, and horizontal detail coefficients. For example, the DWT coefficient HH2 is the matrix equal to the diagonal detail coefficient of the second level of decomposition of Lena.

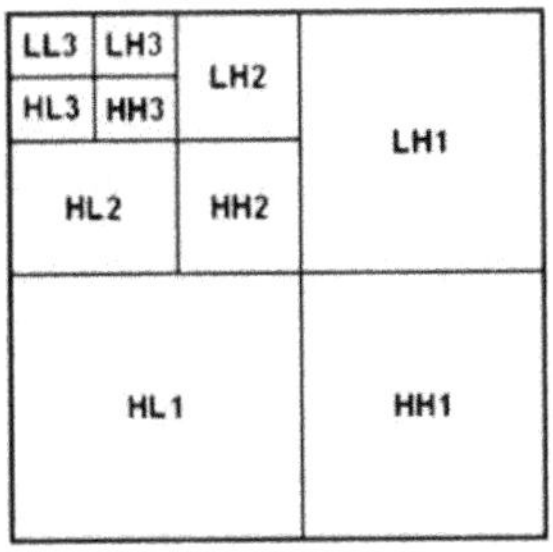

FIGURE 7.4: Wavelets coefficients.

In the example illustrated by Figure 7.5, the watermark is inserted into the diagonal (detail) coefficient HH2, which is a real matrix of size 128×128. To insert the watermark, chaotic iterations are done. The system to iterate is the Boolean vector of size 128^2, constituted by the LSCs of Lena, which are the second least significant bit of each integral value of HH2. Iterate function is the vectorial Boolean negation, and chaotic strategy is defined as in Eq. 7.10, with $U^0 = 1$ and $N = 256^2$.

(a) Original Lena.   (b) Watermarked Lena.

FIGURE 7.5: Data hiding in DWT domain

In this situation, PSNR = 53.45 dB. Pixel values have been modified by at most one level of gray. Mean value of differences is 0.294, when the RMS is equal to 0.542. The alteration can be considered as indistinguishable.

The subject of this chapter is not to propose another watermarking algorithm, but to provide an original theoretical framework to study data hiding schemes under the KMA, KOA, and CMA setups. The algorithm presented above is not proposed as a possible candidate to achieve a robust watermarking (however, [18] and [76] give a first approach of its robustness and its applicability to real-world applications, whereas in [78] its stego-security is established). It is only given here as a second illustrative example of an evaluation of our notion of security.

7.1.5.3   Properties of the chaotic machine

A data hiding scheme which is not topologically secure must not be used in KOA and KMA setups, because it is predictable. However, as stated before, proving that a given data hiding scheme is topologically secure is just the beginning of the study. The next step is to evaluate the quality of its chaotic behavior, by using numerous qualitative and quantitative tools offered by the theory of chaos, as it has been done in Section 7.1.4.2. These tools will allow us to compare spread-spectrum data hiding techniques to *dhCI*, thus helping to decide which scheme can be used in a given setup.

To achieve this goal, qualitative and quantitative chaos properties recalled in this book will be used to evaluate the unpredictability of chaotic iterations $\tilde{G}_{f_0}$. These properties will be possessed by our topologically secure data hiding algorithm, when the switch is chosen.

Qualitative properties

We have proven in Chapter 5 that, as for spread-spectrum, chaotic itera- tions are strongly transitive. Indeed, we have shown in this chapter that these iterations also possess the property of topological mixing (Proposition 7).

The consequence of topological mixing for data hiding are multiple. First, in a situation of topological mixing, topological security can be largely im- proved by considering the number of iterations of the data hiding machine as a secret key. An attacker will reach all of the possible media when iterating without the knowledge of this key. Additionally, he cannot benefit a lot from a KOA setup, by studying media in the neighborhood of the original cover. Moreover, as in a topological mixing situation, it is possible that any hidden message (the initial condition) is sent to the same fixed watermarked content (with different numbers of iterations), and the interest in a KMA setup is drastically reduced. Last, as all the watermarked contents are possible for a given hidden message, depending on the number of iterations, CMA attacks will fail. To sum up, topological mixing and expansiveness are required for a topologically secure data hiding scheme to withstand attacks in KOA and KMA setups, whereas only topological mixing is fundamental in a CMA setup.

Quantitative measures

We have established in Chapter 5 that, like the spread-spectrum, chaotic iterations are sensitive to the initial condition. However, contrary to spread- spectrum techniques, chaotic iterations are expansive.

To sum up, data hiding schemes based on chaotic iterations are topologi- cally secure, with the additional properties of strong transitivity and topolog- ical mixing, as it is in the case of spread-spectrum. However, unlike spread- spectrum, chaotic iteration are expansive (with a constant of expansiveness equal to 1), so chaotic iterations appear to be more secure than spread- spectrum in KOA and KMA setups. The security aspects of *dhCI* data hiding schemes is summed up in Table 7.2.

To extend the comparison between chaotic iterations and spread-spectrum techniques, we need to adapt the proofs detailed in Chapter 5 to use equiv- alent distances on the same phase space as spread-spectrum. This work can be achieved either by reconsidering the modeling of chaotic iterations, or by studying spread-spectrum on the finite set of machine numbers.

7.1.6 Discussion

In this section, our concept of security for data hiding schemes has been presented as a complementary approach to the existing framework. This notion of security, namely the topological security, contributes to the reinforcement of confidence put into existing secure data hiding schemes. Additionally, the study of security in KMA, KOA, and CMA setups is realizable in this context. Finally, this framework can replace stego-security in situations that are not encompassed by it. In particular, this framework is more relevant to give evaluation of data hiding schemes claimed as chaotic.

TABLE 7.2: Evaluation of dhCI scheme

	Property	Switch Mode	Replace Mode
CAPABILITY	Blind watermarking	No	Yes
	Domain	Spatial, frequency	Spatial, frequency
	Authentication	Capable	Capable
	Watermark encryption	Yes	Yes
	Robustness	See [18]	See [18]
	topological security	Switch Mode	Replace Mode
	Devaney's chaos	Yes	?
	Phase space	$\mathcal{X}' = [\![1; \mathsf{N}]\!]^{\mathbb{N}} \times \mathbb{B}^{\mathsf{N}}$	?
	Topology	d'	?
	Qualitative Properties	Switch Mode	Replace Mode
TOPOLOGICAL	Strong transitivity	Yes	?
SECURITY	Topological mixing	Yes	?
	Quantitative Properties	Switch Mode	Replace Mode
	Sensitivity	Yes. $\delta = \mathsf{N} - 1$	?
	expansiveness	Yes. $\varepsilon = 1$	?
		Switch mode	Replace mode
STEGO-SECURITY	Stego-security	Established in [78]	?

In our approach, a data hiding scheme is secure if it is unpredictable. Its iterative process must satisfy the Devaney's chaos property and its level of topological security increases with the number of chaotic properties satisfied by it. This point has been clarified in Sections 7.1.4 and 7.1.5, in which a first study of topological security is proposed using some qualitative and quantitative tools taken from the mathematical theory of chaos.

7.2   Pseudorandom Number Generators

7.2.1   Introduction

Randomness is of importance in many fields such as scientific simulations or cryptography. "Random numbers" can mainly be generated either by a deterministic and reproducible algorithm called a pseudorandom number generator (PRNG), or by a physical non-deterministic process having all the characteristics of a random noise, called a truly random number generator (TRNG). In this section, we focus on reproducible generators, useful for instance in Monte-Carlo based simulators or in several cryptographic schemes.

These domains need PRNGs that are statistically irreproachable. Speed is an important requirement for simulations. On the other side, in cryptography, the great need is to define *secure* generators able to withstand malicious attacks. Roughly speaking, an attacker should not be able in practice to make the distinction between numbers obtained with the secure generator and a true random sequence. In an equivalent formulation, he or she should not be able (in practice) to predict the next bit of the generator, having the knowledge of all the binary digits that have been already released. "Being able in practice" refers here to the possibility to achieve this attack in polynomial time, and to the exponential growth of the difficulty of this challenge when the size of the parameters of the PRNG increases.

Finally, a small part of the community working in this domain focuses on a third requirement, that is, to define chaotic generators. The main idea is to take benefits from a chaotic dynamical system to obtain a generator that is unpredictable, disordered, sensible to its seed, or in other words chaotic. Their desire is to map a given chaotic dynamic into a sequence that seems random and unassailable due to chaos. However, the chaotic maps used as a pattern are defined in the real line whereas computers deal with finite precision numbers. This distortion leads to a deflation of both chaotic properties and speed. Furthermore, authors of such chaotic generators often claim their PRNG as secure due to their chaos properties, but there is no obvious relation between chaos and security as it is understood in cryptography. This is why the use of chaos for PRNG still remains marginal and disputable.

The authors' opinion is that topological properties of disorder, as they have been properly defined in Part II, can reinforce the quality of a PRNG, but they are not substitutable for security or statistical perfection. Indeed, to the authors' minds, such properties can be useful in the two following situations. On the one hand, a post-treatment based on a chaotic dynamical system can be applied to a PRNG statistically deflective, in order to improve its statistical properties. Such an improvement can be found, for instance, in [9,26]. On the other hand, chaos can be added to a fast, statistically perfect PRNG and/or a cryptographically secure one, in a case where chaos can be of interest, *only if these last properties are not lost during the proposed post-treatment*. Such an assumption is behind the work presented in this section. It leads to the attempts to define a family of PRNGs that are chaotic while being fast and statistically perfect, or cryptographically secure. Let us finish this paragraph by noticing that, in this section, statistical perfection refers to the ability to pass the whole *BigCrush* battery of tests, which is widely considered as the most stringent statistical evaluation of a sequence claimed as random. This battery can be found in the well-known TestU01 package [90].

Let us first introduce the notion of security for PRNGs, as defined by the cryptography community.

7.2.2 Secured generators

The standard definition of *indistinguishability* used is the classical one as defined for instance in [72, Chapter 3]. This property shows that predicting the future results of the PRNG cannot be done in a reasonable time compared to the generation time. It is important to emphasize that this is a relative notion between breaking time and the sizes of the keys/seeds. Of course, if small keys or seeds are chosen, the system can be broken in practice. But it also means that if the keys/seeds are large enough, the system is secured.

In this section the concatenation of two strings u and v is classically denoted by uv. In a cryptographic context, a pseudorandom generator is a deterministic algorithm G transforming strings into strings and such that, for any seed s of length m, $G(s)$ (the output of G on the input s) has size $\ell_G(m)$ with $\ell_G(m) > m$. The notion of *secure* PRNGs can now be defined as follows.

Definition 70 A cryptographic PRNG G is secure if for any probabilistic polynomial time algorithm D, for any positive polynomial p, and for all sufficiently large m's,

$$|\Pr[D(G(U_m)) = 1] - Pr[D(U_{\ell_G(m)}) = 1]| < \frac{1}{p(m)},$$

where U_r is the uniform distribution over $\{0,1\}^r$ and the probabilities are taken over U_m, $U_{\ell_G(m)}$ as well as over the internal coin tosses of D.

As stated before, it means that there is no polynomial time algorithm that can distinguish a perfect uniform random generator from G with a non negligible probability. An equivalent formulation of this well-known security property means that it is possible *in practice* to predict the next bit of the generator, knowing all the previously produced ones. The interested reader is referred to [72, Chapter 3] for more information. Note that it is quite easily possible to change the function ℓ into any polynomial function ℓ' satisfying $\ell'(m) > m$) [72, Chapter 3.3].

7.2.3 Pseudorandom generators based on chaotic iterations (CI PRNG)

We have proposed in [11–15, 26–28, 136] a post-treatment on PRNGs making them behave as a chaotic dynamical system. Such a post-treatment leads to a new category of PRNGs. We have shown that proofs of Devaney's chaos can be established for this family, and that the sequence obtained after this post-treatment can pass the NIST [36], DieHard [103], and TestU01 [90] batteries of tests, even if the inputted generators cannot. Finally, we have recently shown that this post-treatment leads to a secured PRNG when one of the inputted generators is cryptographically secure.

The proposition of the remainder of this section is to summarize these works.

7.2.3.1 Some well-known pseudorandom generators

Let us first introduce three existing generators.
The logistic map defined by

$$x^{n+1} = \mu \, x^n (1 - x^n), \qquad (7.11)$$

has been introduced in the first part of this book. In 1947, Ulam and Von Neumann [132] studied the possibility to use it as a PRNG. To do so, it only suffices to map the states of the system $(x^n)_{n \in \mathbb{N}}$ on $\{0,1\}^{\mathbb{N}}$. An easy way to realize this transformation, that is to translate x^n into a binary digit r, consists in using a threshold value like in Algorithm 1 (this threshold is here equal to 0.5). Another way to obtain an integer sequence starting with a real dynamical system consists in operating a shift function on the decimal part of each obtained real number, and to return at each iteration the integral part, as in Algorithm 2. It seems very difficult to produce something secure using the logistic map (see, *e.g.*, [7]). Indeed, this function possesses various biases that can potentially be exploited by an adversary [86].

Algorithm 1 An arbitrary iteration of a PRNG based on the logistic map (version 1)

Input: internal state x (a real number)
Output: r (one bit)
1: $x \leftarrow 4x(1 - x)$
2: **if** $x < 0.5$ **then**
3:     $r \leftarrow 0$;
4: **else**
5:     $r \leftarrow 1$;
6: **end if**
7: return r

Algorithm 2 An arbitrary iteration of a PRNG based on the logistic map (version 2)

Input: internal state x (a real number)
Output: r (an integer)
1: $x \leftarrow 4x(1 - x)$
2: $r \leftarrow \lfloor 10000000x \rfloor$
3: return r

XORshift is another category of very fast but insecure PRNGs formerly proposed by George Marsaglia [104]. These PRNGs consist of the iteration of the *exclusive or* (XOR) operation between the previously obtained state and a circular shifted version of it.

More precisely, the internal state of a XORshift is a binary vector, and at each iteration, the new state is computed by applying a predefined number of operations of the "XORshift kind" on blocks of w bits of the current state, with $w = 32$ or 64. These operations are defined as follows: the w bits block is replaced by a *bitwise exclusive or* between this block and a version shifted of a bits, on the left or on the right, where $0 < a < w$ (see Algorithm 3).

Algorithm 3 An arbitrary round of XORshift algorithm

Input: the internal state z (a 32-bits word)
Output: y (a 32-bits word)
1: $z \leftarrow z \oplus (z \ll 13)$;
2: $z \leftarrow z \oplus (z \gg 17)$;
3: $z \leftarrow z \oplus (z \ll 5)$;
4: $y \leftarrow z$;
5: return y

XORshift of the Algorithm 3 has a period equal to $2^{32} - 1 \approx 4,29 \times 10^9$. This period can be enlarged using improved versions of this PRNG. However, as stated before, the XORshift is not cryptographically secure.

On the contrary, ISAAC is a secured PRNG proposed by Robert Jenkins in 1996 [84]. The name of this PRNG is an acronym for "Indirection, Shift, Accumulate, Add and Count". This algorithm is quite close to the famous RC4 [109]. ISAAC uses as internal state x an array of 256 integers of 32 bits (*c.f.* Algorithm 4). Results are recorded in another array of 256 integers, denoted by r. r will be returned and will replace x when each word of x will be used. In Algorithm 4, variables a, b, and c have been initialized at the first iteration ($a = b = c = 0$), and they are recomputed at each iteration. Finally, the value $f(a, i)$ is a word of 32 bits, defined for all a and i into $\{0, \ldots, 255\}$ in the following manner:

$$f(a, i) = \begin{cases} a \ll 13 & \text{if } i \equiv 0 \ mod \ 4, \\ a \gg 6 & \text{if } i \equiv 1 \ mod \ 4, \\ a \ll 2 & \text{if } i \equiv 2 \ mod \ 4, \\ a \gg 16 & \text{if } i \equiv 3 \ mod \ 4. \end{cases} \tag{7.12}$$

As this algorithm realizes only 19 operations for each output of 32 bits, it is extremely fast on 32 bits computers.

7.2.3.2   The "Old CI" generator: algorithms and examples

We have introduced everything required to be able now to define our first generator.

Chaotic iterations used as a PRNG

The first generator we have proposed, which is denoted by CI(PRNG1,PRNG2), has been constructed in the following manner. Let us consider two pseudorandom generators, namely PRNG1 and PRNG2.

Algorithm 4 An arbitrary round of ISAAC

Inputs: a, b, c and the internal state x
Output: a vector r having 256 words of 32 bits

1: $c \leftarrow c + 1$;
2: $b \leftarrow b + c$;
3: **while** $i = 0, \ldots, 255$ **do**
4:     $s \leftarrow x_i$;
5:     $a \leftarrow f(a, i) + x_{(i+128) \bmod 256}$;
6:     $x_i \leftarrow a + b + x_{(x \gg 2) \bmod 256}$;
7:     $r_i \leftarrow s + x_{(x_i \gg 10) \bmod 256}$;
8:     $b \leftarrow r_i$;
9: **end while**
10: return r

- Chaotic iterations are realized using the vectorial negation and the PRNG1 as strategy.

- We do not publish all the states of the system, but only the ones selected by PRNG2.

The objectives are:

1. To run approximately at the same speed as PRNG1 and PRNG2.

2. To possess chaos properties, even if the inputted PRNGs do not have these properties.

3. To improve the results against statistical tests (thanks to the chaotic properties).

4. To preserve the cryptographically secure character, when one of the two inputted generator has such a property.

More precisely, let $N \in \mathbb{N}^*, N \geqslant 2$. The initial state $x^0 \in \mathbb{B}^N$ is a vector of bits obtained with the seed provided by the user, and the chaotic strategy $(S^n)_{n \in \mathbb{N}} \in [\![1, N]\!]^{\mathbb{N}}$ is the PRNG1 (initialized using another part of the provided seed). The iteration function f is the vectorial negation[1]:

$$f_0 : (x_1, ..., x_N) \in \mathbb{B}^N \longmapsto (\overline{x_1}, ..., \overline{x_N}) \in \mathbb{B}^N. \tag{7.13}$$

At each iteration, only the coordinate S^i of the state x^n is updated, as follows:

$$x_i^n = \begin{cases} x_i^{n-1} & \text{if } i \neq S^i, \\ \\ \overline{x_i^{n-1}} & \text{if } i = S^i. \end{cases} \tag{7.14}$$

---

[1]Other Boolean functions can be imagined; the sole requirement is to make the CIs chaotic.

Finally, some of these vectors are randomly extracted, and their coordinates constitute the bit flow of our pseudorandom generator. To do so, a finite nonempty subset $\mathcal{M}$ of $\mathbb{N}^*$ is introduced, and a sequence $(m^n)_{n\in\mathbb{N}} \in \mathcal{M}^\mathbb{N}$ is generated using PRNG2. The generator returns the followed values: the coordinates of x^{m^0}, followed by the coordinates of $x^{m^0+m^1}$, following by those of $x^{m^0+m^1+m^2}$, *etc.* In other words, the generator returns the following bits:

$$x_1^{m_0} x_2^{m_0} x_3^{m_0} \ldots x_N^{m_0} x_1^{m_0+m_1} x_2^{m_0+m_1} \ldots x_N^{m_0+m_1} x_1^{m_0+m_1+m_2} x_2^{m_0+m_1+m_2} \ldots$$

(7.15)

or the following integers:

$$x^{m_0} x^{m_0+m_1} x^{m_0+m_1+m_2} \ldots$$

The basic design procedure of the novel generator, initially published in [26, 28, 34], can indeed be summarized as in Algorithm 5. The internal state is x, the output array is r. a and b are those computed by PRNG1 and PRNG2. Finally, k and N are constants, and $\mathcal{M}$ is equal to $\{k, k+1\}$, with $k \geqslant 3N$. Some explanations about the choices defining this old CI PRNG are given thereafter.

Algorithm 5 An arbitrary round of the old CI generator

Input: the internal state x (an array of N 1-bit words)
Output: an array r of N 1-bit words
1: $a \leftarrow PRNG1()$;
2: $m \leftarrow a \bmod 2 + k$;
3: **while** $i = 0, \ldots, m$ **do**
4:    $b \leftarrow PRNG2()$;
5:    $S \leftarrow b \bmod N$;
6:    $x_S \leftarrow \overline{x_S}$;
7: **end while**
8: $r \leftarrow x$;
9: **return** r;

A few words on the seed

The initial state of the system x^0 and the first term y^0 of the generator given as input can be initialized either by using the system clock (in seconds since the Epoch[2]), or by a number given by the user. Various ways to proceed are obviously possible, to obtain the seed by using the time. For instance, let t be the decimal part of the current time. Then x^0 can be t (mod 2^N) in binary digits, and $y^0 = t$.

In the framework of cryptographically secure generators and without the seed, the adversary should not be able to make a correct prediction on the

---

[2]The epoch represents the initial date from which the time on current exploitation systems is measured. This date depends on the system.

produced bits, even when all the other details of the generator are known [131]. In the best situation, these seeds should not present any bias, that is, should be really random. Such seeds can be obtained, for instance, using a "physical noise," like the timing keystrokes of the user.

Definition of the set $\mathcal{M}$

The logic that has led the authors to define the generator as it has been presented in the previous section is easily understandable: chaotic iterations have good topological properties of disorder and they only manipulate integers. Thus these properties are preserved on finite machines (integrally if the strategy is a stream obtained from the outside world), and this generator is fast, because it only uses integers and the vectorial negation. So the objectives enumerated above seems to be attainable. In that situation, the role played by the second inputted PRNG is not clear. Additionally the number of terms that must be computed at each iteration is enlarged: all the terms of the PRNG2 must be obtained. Similarly the number of states to obtain between two outputs increases, deflating the speed of the proposed generator by doing so.

Indeed, reasons requiring this complexity can be illustrated in the following manner. It is impossible to predict with an high precision and certitude what the weather will be in one month, but it is easy to predict what it will be in two minutes from the current weather. In other words, chaos needs time. In this first version of generator, only one bit is swapped at each iteration in a vector of size N. Such an operation must obviously be realized a certain number of times between two outputs. This first idea is the foundation of this first PRNG, which is only a proof of concept.

Thus values that can be taken by m^i (that is, the set $\mathcal{M}$) should be defined carefully. The first idea consists of taking the m^i constant, which has the advantage of reducing the inputted pseudorandom generators to PRNG1. In other words: take a singleton of the form $\mathcal{M} = \{k\}, k > 1$, and thus only output the states $x^{kn}, n \in \mathbb{N}$.

This method has the advantage of preventing the computations related to the PRNG2. However, the following example suffices to show that this method is not suitable. Let us consider a system having 4 states ($\mathsf{N} = 4$), and suppose that for a given i, the generator produces $x^i = (0,0,0,0)$ as output. Table 7.3 contains the possible values for x^{i+1} with various choices of singletons $\mathcal{M}$. As may be seen, each choice of $\mathcal{M}$ restricts the set of possible values x^{i+1}. This fact can be easily understood by considering that the negation is an involution: $\bar{\bar{x}} = x$. It can be remarked too that sets of the form $\mathcal{M} = \{k, k+1\}, k \geqslant 3$ may fit, whereas $\{1, 2, 3, 4\}$ should lead to non uniform outputs.

The first approach to determine $\mathcal{M}$ can be an experimental one: histograms of successive values produced by the generator, depending on the choice of $\mathcal{M}$, can be plotted. These histograms are constructed in the following manner. z is initialized to the null matrix of size 15×15. For each couple (x, y) in $[\![1, 15]\!]$, and for all i lower than 10^6, if $x^i = x$ and $x^{i+1} = y$, then $z_{x,y}$ is incremented of 1. As after each value produced by the generator at the i-th iterate, all

TABLE 7.3: Possible values for x^{i+1} for various singletons $\mathcal{M}$

x^{i+1}	$\mathcal{M}:$ {1}	{2}	{3}	{4}	{5}	{6}	{7}	{8}	{9}	{10}	{11}	{12}	{13}
0		√		√		√		√		√		√	
1	√		√		√		√		√		√		√
2	√		√		√		√		√		√		√
3		√		√		√		√		√		√	
4	√		√		√		√		√		√		√
5		√		√		√		√		√		√	
6		√		√		√		√		√		√	
7			√		√		√		√		√		√
8	√		√		√		√		√		√		√
9		√		√		√		√		√		√	
10		√		√		√		√		√		√	
11			√		√		√		√		√		√
12		√		√		√		√		√		√	
13			√		√		√		√		√		√
14			√		√		√		√		√		√
15				√		√		√		√		√	

the values in $[\![1,15]\!]$ must be attainable with the same probability for x^{i+1}, in order to obtain a generator able to pass statistical tests, these histograms must be as flat as possible.

Examples of experimental results are given in Figures 7.6a, 7.6b, 7.6c, 7.6d, 7.6e, 7.6f, and 7.6g. They allow us to understand why we have chosen, for this proof of concept, to use $\mathcal{M} = \{k, k+1\}$ with $k \geqslant 3N$.

Let us now give an example of run for the proposed generator.

Example

In this example, $N = 5$ and $\mathcal{M} = \{4,5\}$ are chosen for easy understanding. The initial state of the system x^0 can be seeded by the decimal part of the current time. For example, the current time in seconds since the Epoch is 1237632934.484084, so $t = 484084$. And thus $x^0 = t \pmod{32}$, which must be written in binary digits, so $x^0 = (1,0,1,0,0)$. m and S can now be computed from PRNG1 and PRNG2. Suppose that we have obtained:

- $m = 4, 5, 4, 4, 4, 4, 5, 5, 5, 5, 4, 5, 4,...$

- $S = 2, 4, 2, 2, 5, 1, 1, 5, 5, 3, 2, 3, 3,...$

Chaotic iterations are done with initial state x^0, vectorial logical negation f_0,

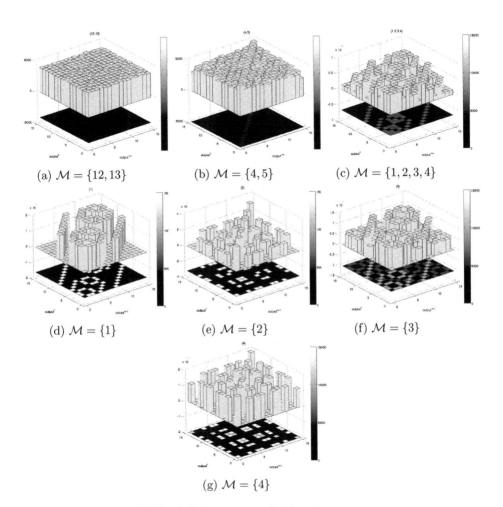

(a) $\mathcal{M} = \{12, 13\}$ (b) $\mathcal{M} = \{4, 5\}$ (c) $\mathcal{M} = \{1, 2, 3, 4\}$

(d) $\mathcal{M} = \{1\}$ (e) $\mathcal{M} = \{2\}$ (f) $\mathcal{M} = \{3\}$

(g) $\mathcal{M} = \{4\}$

FIGURE 7.6: Histogram and intensity maps

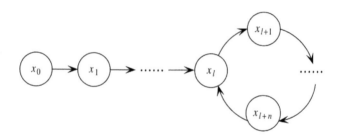

FIGURE 7.7: A pseudo orbit of a digital chaotic system

and strategy S. The result is presented in Table 7.6. Let us recall that sequence m gives the states x^n to return: $x^4, x^{4+5}, x^{4+5+4}, \ldots$

TABLE 7.4: Application example

$m:$	4				x^4	5					x^9	4				x^{13}
S	2	4	2	2		5	1	1	5	5		3	2	3	3	
x^0																
1					1	$\xrightarrow{1}0\xrightarrow{1}1$					1					1
0	$\xrightarrow{2}1$	$\xrightarrow{2}0\xrightarrow{2}11$			1						1	$\xrightarrow{2}0$				0
1					1						1	$\xrightarrow{3}0$	$\xrightarrow{3}1\xrightarrow{3}00$			0
0	$\xrightarrow{4}1$				1						1					1
0					0	$\xrightarrow{5}1$		$\xrightarrow{5}0\xrightarrow{5}11$								1

Binary Output:
$$x_1^0 x_2^0 x_3^0 x_4^0 x_5^0 x_1^4 x_2^4 x_3^4 x_4^4 x_5^4 x_1^9 x_2^9 x_3^9 x_4^9 x_5^9 x_1^{13} x_2^{13}\ldots = 10100111101111110\ldots$$
Integer Output: $x^0, x^0, x^4, x^6, x^8 \ldots = 20, 30, 31, 19\ldots$

So, in this example, the generated binary digits are: 10100111101111110011... Or the integers are: 20, 30, 31, 19... Let us finally evaluate the period of this first PRNG.

On the periodicity of chaotic orbit

Since chaotic iterations are constrained in a discrete space with 2^N elements, it is obvious that every chaotic orbit will eventually be periodic, i.e., finally goes to a cycle with limited length not greater than 2^N. The schematic view of a typical orbit of a digital chaotic system is shown in Figure 7.7. Generally speaking, each digital chaotic orbit includes two connected parts: $x^0, x^1, \ldots, x^{l-1}$ and $x^l, x^{l+1}, \ldots, x^{l+n}$, which are respectively called transient (branch) and cycle. Accordingly, l and $n+1$ are respectively called transient length and cycle period, and $l+n$ is called orbit length. Thus,

TABLE 7.5: Ideal cycle period

PRNG		Ideal cycle
Logistic map		∞
XORshift		$2^{32} - 1$
ISAAC		2^{8295}
Old CI algorithms	**Logistic map 1+Logistic map 2**	∞
	XORshift+XORshift	2^{65}
	XORshift+ISAAC	2^{8328}
	ISAAC+ISAAC	2^{16591}

Definition 71 A sequence $x = (x^1, ..., x^n)$ is said to be cyclic if a subset of successive terms is repeated from a given rank, until the end of x.

This generator based on discrete chaotic iterations generated by two pseudorandom sequences (m and w) has a long cycle length. It is easy to check that, if the cycle period of m and w are respectivelly n_m and n_w, then in an ideal situation, the cycle period of the novel sequence is $n_m \times n_w \times 2$ (because $\bar{\bar{x}} = x$). Table 7.5 gives the ideal cycle period of various generators.

7.2.3.3 The "New CI" algorithm

To improve the Old CI PRNG and to make it really usable in practice, it is possible to process as detailed on the following pages.

A similar method is followed: chaotic iterations are realized to generate a sequence $(x^n)_{n \in \mathbb{N}} \in \left(\mathbb{B}^N\right)^{\mathbb{N}}$ ($N \in \mathbb{N}^*, N \geqslant 2$) of vectors of bits, that correspond to the successive states of the iterated system. Once again, some of these vectors will be randomly extracted, and the pseudorandom binary sequence will be their coordinates. However, we will now prevent useless operations. To do so, we will proceed as follows.

Initial state $x^0 \in \mathbb{B}^N$ is a Boolean vector taken as a seed (see Section 7.2.3.2) and chaotic strategy $(S^n)_{n \in \mathbb{N}} \in [\![1, N]\!]^{\mathbb{N}}$ is an irregular decimation of a random number sequence (Section 7.2.3.3). The iterate function f is the vectorial Boolean negation:

$$f_0 : (x_1, ..., x_N) \in \mathbb{B}^N \longmapsto (\overline{x_1}, ..., \overline{x_N}) \in \mathbb{B}^N.$$

At each iteration, only the S^i-th component of state x^n is updated, as follows: $x_i^n = x_i^{n-1}$ if $i \neq S^i$, else $x_i^n = \overline{x_i^{n-1}}$. Finally, some x^n are selected by a sequence m^n as the pseudorandom bit sequence of our generator. $(m^n)_{n \in \mathbb{N}} \in \mathcal{M}^{\mathbb{N}}$ is computed from a PRNG, such as XORshift sequence $(y^n)_{n \in \mathbb{N}} \in [\![0, 2^{32} - 1]\!]$ (see Section 7.2.3.3). So, the generator returns the following values:

$$x_1^{m_0} x_2^{m_0} x_3^{m_0} \ldots x_N^{m_0} x_1^{m_0+m_1} x_2^{m_0+m_1} \ldots x_N^{m_0+m_1} x_1^{m_0+m_1+m_2} \ldots$$

or states:

$$x^{m_0} x^{m_0+m_1} x^{m_0+m_1+m_2} \ldots$$

Indeed the principal problem with the Old CI generator is that a same bit can be swapped twice between two outputs, which is useless loss of time. By doing so, it is a necessity to iterate a sufficient number of times to prevent deflated statistics. The issue consists of picking the right strategy, in such a way that each bit of the system is updated at most one time between two outputs. Such a desire is at the origin of the idea of decimation in $\mathcal{S}$, which will be detailed below. Furthermore, it is a necessity to redefine the sequence m^n, which will now designate the *number of bits that must change between two outputs*. This redefinition, useful in the realization of the decimation in $\mathcal{S}$, leads to a subtlety presented in the next paragraph.

Sequence m of returned states

The output of the sequence (y^n) is uniform in $[\![0, 2^{32}-1]\!]$. However, we do not want the output of (m^n) to be uniform in $[\![0, N]\!]$, because in this case, the returns of our generator will not be uniform in $[\![0, 2^N-1]\!]$, as it is illustrated in the following example.

Let us suppose that $x^0 = (0,0,0)$. Then $m^0 \in [\![0, 3]\!]$.

- If $m^0 = 0$, then no bit will change between the first and the second output of our new CI PRNG. Thus $x^1 = (0,0,0)$.

- If $m^0 = 1$, then exactly one bit will change, which leads to three possible values for x^1, namely $(1,0,0)$, $(0,1,0)$, and $(0,0,1)$.

- etc.

As each value in $[\![0, 2^3-1]\!]$ must be returned with the same probability, then the values $(0,0,0)$, $(1,0,0)$, $(0,1,0)$, and $(0,0,1)$ must occur for x^1 with the same frequency. We see that, in this example, $m^0 = 1$ must be three times more probable than $m^0 = 0$. This assessment leads to the following general definition for the probability of $m = i$:

$$P(m^n = i) = \frac{C_N^i}{2^N}. \tag{7.16}$$

These findings lead to define the selection function g_1:

$$m^n = g_1(y^n) = \begin{cases} 0 \text{ if } 0 \leqslant \frac{y^n}{2^{32}} < \frac{C_N^0}{2^N}, \\ 1 \text{ if } \frac{C_N^0}{2^N} \leqslant \frac{y^n}{2^{32}} < \sum_{i=0}^{1} \frac{C_N^i}{2^N}, \\ 2 \text{ if } \sum_{i=0}^{1} \frac{C_N^i}{2^N} \leqslant \frac{y^n}{2^{32}} < \sum_{i=0}^{2} \frac{C_N^i}{2^N}, \\ \vdots \quad \vdots \\ N \text{ if } \sum_{i=0}^{N-1} \frac{C_N^i}{2^N} \leqslant \frac{y^n}{2^{32}} < 1. \end{cases} \tag{7.17}$$

Using such a function, it is possible to transform uniformly distributed sequence y^n of $[\![0, 2^{32} - 1]\!]$ to an another sequence m^n belonging to $[\![0, N]\!]$, which has a distribution adequate with the New CI PRNG. Such a function can be easily adapted to other intervals than $[\![0, 2^{32} - 1]\!]$, if required. Remark that this function is not unique; for instance, the function $g_2()$ defined below works too:

$$m^n = g_2(y^n) = \begin{cases} N \text{ if } 0 \leqslant \frac{y^n}{2^{32}} < \frac{C_N^0}{2^N}, \\ N-1 \text{ if } \frac{C_N^0}{2^N} \leqslant \frac{y^n}{2^{32}} < \sum_{i=0}^{1} \frac{C_N^i}{2^N}, \\ N-2 \text{ if } \sum_{i=0}^{1} \frac{C_N^i}{2^N} \leqslant \frac{y^n}{2^{32}} < \sum_{i=0}^{2} \frac{C_N^i}{2^N}, \\ \vdots \quad \vdots \\ 0 \text{ if } \sum_{i=0}^{N-1} \frac{C_N^i}{2^N} \leqslant \frac{y^n}{2^{32}} < 1. \end{cases} \tag{7.18}$$

In order to evaluate the proposed method and compare its statistical properties with various other methods, the density histogram and intensity map of adjacent output have first been computed. The length of x is $N = 4$ bits, and the initial conditions and control parameters are the same. A large number of sampled values are simulated (10^6 samples). Figure 7.8a and Figure 7.8b show the intensity map for $m^n = g_1(y^n)$ and $m^n = g_2(y^n)$ respectively. Uniform histograms and flat color intensity maps are obtained when using the proposed schemes. Another illustration is given in Figure. 7.8c whereas its uniformity is further justified by the tests presented in Section 7.2.4.4.

The chaotic strategy

The chaotic strategy $(S^k) \in [\![1, N]\!]^{\mathbb{N}}$ is generated from a second XOR-shift sequence $(b^k) \in [\![1, N]\!]^{\mathbb{N}}$. The only difference between the sequences S and b is that some terms of b are discarded, in such a way that $\forall k \in \mathbb{N}, (S^{M^k}, S^{M^k+1}, \ldots, S^{M^{k+1}-1})$ does not contain any given integer twice, where $M^k = \sum_{i=0}^{k} m^i$. Therefore, no bit will change more than once between two successive outputs of this PRNG, increasing the speed of the former generator by doing so. S is said to be "an irregular decimation" of b. This decimation can be obtained by the following process.

Let $(d^1, d^2, \ldots, d^N) \in \{0,1\}^N$ be a marked sequence, such that whenever $\sum_{i=1}^{N} d^i = m^k$, then $\forall i, d_i = 0$ ($\forall k$, the sequence is reset when d contains m^k times the number 1). This marked sequence will control the XORshift b as follows:

- if $d^{b^j} \neq 1$, then $S^k = b^j$, $d^{b^j} = 1$, and $k = k + 1$,

- if $d^{b^j} = 1$, then b^j is discarded.

For example, if $b = 142\underline{2}334142\underline{1}12234...$ and $m = 4341...$, then $S = 1423\ 341\ 4123\ 4...$

If we do not use the mark sequence, then one position may change more than once and the balance property will not be checked, due to the fact

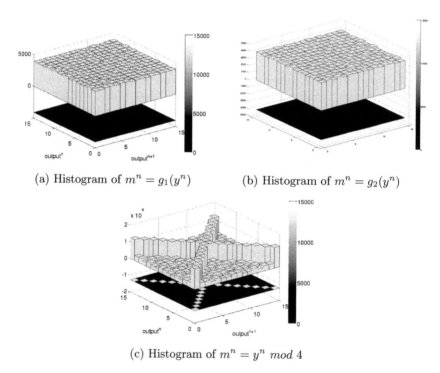

(a) Histogram of $m^n = g_1(y^n)$ (b) Histogram of $m^n = g_2(y^n)$

(c) Histogram of $m^n = y^n \bmod 4$

FIGURE 7.8: Histogram and intensity maps

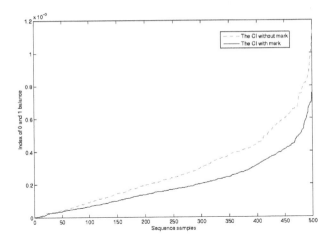

FIGURE 7.9: Balance property

that $\bar{\bar{x}} = x$. As an example, for b and m as in the previous example, $S = 1422\ 334\ 1421\ 1...$ and $S = 14\ 4\ 42\ 1...$ lead to the same output.

To check the balance property, a set of 500 sequences are generated with and without decimation, each sequence containing 10^6 bits. Figure 7.9 shows the percentages of differences between zeros and ones, and presents a better balance property for the sequences with decimation. This claim will be verified in the tests section (Section 7.2.4.4). Another example is given in Table 7.6, in which r means "reset" and the integers that are underlined in sequence b are discarded.

It is now possible to give the algorithm of the optimized generator.

New CI(XORshift, XORshift) Algorithm

The basic design procedure of the novel generator is summed up in Algorithm 6. The internal state is x, the output state is r. a and b are those computed by the two PRNGs provided as inputs. The value $g_1(a)$ is an integer as given in Eq. 7.17. Last, N is a constant defined by the user. It can be compared with the basic design procedure of the old generator detailed in Algorithm 5.

To close this presentation, a complete example of number generation using the New CI PRNG is given below.

Illustrative Example

As previously, N = 4 and the initial state of the system x^0 is seeded by the decimal part t of the current time. Suppose that: $x^0 = (0, 1, 0, 0)$.

To compute the m sequence, Equation 7.17 can be adapted to this example

---

Algorithm 6 An arbitrary round of the new CI generator

Input: the internal state x (N bits)
Output: a state r of N bits

1: **for** $i = 0, \ldots, N$ **do**
2:     $d_i \leftarrow 0$
3: **end for**
4: $a \leftarrow PRNG1()$
5: $m \leftarrow f(a)$
6: $k \leftarrow m$
7: **while** $i = 0, \ldots, k$ **do**
8:     $b \leftarrow PRNG2() \bmod \mathsf{N}$
9:     $S \leftarrow b$
10:     **if** $d_S = 0$ **then**
11:         $x_S \leftarrow \overline{x_S}$
12:         $d_S \leftarrow 1$
13:     **else if** $d_S = 1$ **then**
14:         $k \leftarrow k + 1$
15:     **end if**
16: **end while**
17: $r \leftarrow x$
18: **return** r

---

as follows:

$$m^n = g_1(y^n) = \begin{cases} 0 & \text{if } 0 \leqslant \frac{y^n}{2^{32}} < \frac{1}{16}, \\ 1 & \text{if } \frac{1}{16} \leqslant \frac{y^n}{2^{32}} < \frac{5}{16}, \\ 2 & \text{if } \frac{5}{16} \leqslant \frac{y^n}{2^{32}} < \frac{11}{16}, \\ 3 & \text{if } \frac{11}{16} \leqslant \frac{y^n}{2^{32}} < \frac{15}{16}, \\ 4 & \text{if } \frac{15}{16} \leqslant \frac{y^n}{2^{32}} < 1, \end{cases} \qquad (7.19)$$

where y is generated by a XORshift seeded with the current time. We can see that the probabilities of occurrences of $m = 0$, $m = 1$, $m = 2$, $m = 3$, $m = 4$, are $\frac{1}{16}, \frac{4}{16}, \frac{6}{16}, \frac{4}{16}, \frac{1}{16}$, respectively. Let us recall that this m determines the next output x. For instance,

- If $m = 0$, the following x will be $(0, 1, 0, 0)$.

- If $m = 1$, x^1 can be either $(1, 1, 0, 0)$, $(0, 0, 0, 0)$, $(0, 1, 1, 0)$, or $(0, 1, 0, 1)$.

- If $m = 2$, the following x can be $(1, 0, 0, 0)$, $(1, 1, 1, 0)$, $(1, 1, 0, 1)$, $(0, 0, 1, 0)$, $(0, 0, 0, 1)$, or $(0, 1, 1, 1)$.

- If $m = 3$, the following x can be $(0, 0, 1, 1)$, $(1, 1, 1, 1)$, $(1, 0, 0, 1)$, or $(1, 0, 1, 0)$.

TABLE 7.6: Example of New CI(XORshift,XORshift) generation

m	4					2		2		
k	4	+1				2		2	+1	
b	1	4	2	2	3	3	4	1	1	4
d	r $\begin{pmatrix}1\\0\\0\\0\end{pmatrix}\begin{pmatrix}1\\0\\0\\1\end{pmatrix}\begin{pmatrix}1\\1\\0\\1\end{pmatrix}\begin{pmatrix}1\\1\\1\\1\end{pmatrix}$					r $\begin{pmatrix}0\\0\\1\\0\end{pmatrix}\begin{pmatrix}0\\0\\1\\1\end{pmatrix}$		r $\begin{pmatrix}1\\0\\0\\0\end{pmatrix}\begin{pmatrix}1\\0\\0\\1\end{pmatrix}$		
S	1	4	2	3		3	4	1	4	

x^0				x^4		x^6				x^8
0	$\xrightarrow{1}1$			1		1	$\xrightarrow{1}0$			0
1		$\xrightarrow{2}0$		0		0				0
0			$\xrightarrow{3}1$ 1	1	$\xrightarrow{3}0$	0				0
0	$\xrightarrow{4}1$			1		$\xrightarrow{4}0$ 0	0		$\xrightarrow{4}1$ 1	1

Output in bits: $x_1^0 x_2^0 x_3^0 x_4^0 x_1^4 x_2^4 x_3^4 x_4^4 x_1^6 x_2^6 ... = 0100101110000001...$
Output in integers: $x^0, x^4, x^6, x^8 ... = 4, 11, 8, 1...$

- Finally, if $m = 4$, then $x^1 = (1, 0, 1, 1)$.

In the simulation we realized as example, $m = 0, 4, 2, 2, 3, 4, 1, 1, 2, 3, 0, 1, 4, ...$ b is computed with a XORshift generator too, but with another seed. We have found $b = 1, 4, 2, 2, 3, 3, 4, 1, 1, 4, 3, 2, 1, ...$

Chaotic iterations are made with initial state x^0, vectorial logical negation f_0, and strategy S. The result is presented in Table 7.6. Let us recall that sequence m gives the states x^n to return, which are here $x^0, x^{0+4}, x^{0+4+2}, ...$ So, in this example, the output of the generator is: $10100111101111110011...$ or $4,4,11,8,1...$

7.2.4 Experimental study of the randomness of the proposed generators

A theoretical proof for the randomness of a generator is impossible to give, therefore statistical inference based on observed sample sequences produced by the generator seems to be a good option. Considering the properties of binary random sequences, various statistical tests can be designed to evaluate the assertion that the sequence is generated by a random source. We have performed some statistical tests for the two CI PRNGs we proposed. These tests include TestU01 [127], NIST suite [119], DieHard battery of tests [103], and some Comparative test parameters. For completeness of this book and for reference, we give in the remainder of this section a brief description of each of the aforementioned tests.

7.2.4.1 Some well-known statistical batteries of tests for PRNGs

The NIST battery of tests

Among the numerous standard tests for pseudorandomness, a convincing way to show the randomness of the produced sequences is to confront them with the NIST (National Institute of Standards and Technology) Statistical Test, because it is an up-to-date test suite proposed by the Information Technology Laboratory (ITL). A new version of the Statistical Test Suite (Version 2.0) was released in August 11, 2010.

The NIST test suite SP 800-22 is a statistical package consisting of 15 tests. They were developed to test the randomness of binary sequences produced by hardware or software based cryptographic PRNGs. These tests focus on a variety of different types of non-randomness that could exist in a sequence.

For each statistical test, a set of $p-values$ (corresponding to the set of sequences) is produced. The interpretation of empirical results can be conducted in a various number of ways. In this book, the examination of the distribution of $p-values$ to check for uniformity ($p-value_T$) is used. The distribution of $p-values$ is examined to ensure uniformity. If $p-value_T \geqslant 0.0001$, then the sequences can be considered to be uniformly distributed.

In our experiments, 100 sequences (s = 100), each with 1,000,000-bit long, are generated and tested. If the $p-value_T$ of any test is smaller than 0.0001, the sequences are considered to be not good enough and the generating algorithm is not suitable for usage.

In what follows, the fifteen tests of the NIST Statistical test suite are recalled. A more detailed description for these tests can be found in [119].

- **Frequency (Monobit) Test (FT)** is to determine whether the number of ones and zeros in a sequence are approximately the same as would be expected for a truly random sequence.

- **Frequency Test within a Block (FBT)** is to determine whether the frequency of ones in a M-bit block is approximately $M/2$, as would be expected under an assumption of randomness (M is the length of each block).

- **Runs Test (RT)** is to determine whether the number of runs of ones and zeros of various lengths is as expected for a random sequence. In particular, this test determines whether the oscillation between such zeros and ones is too fast or too slow.

- **Test for the Longest Run of Ones in a Block (LROBT)** is to determine whether the length of the longest run of ones within the tested sequence is consistent with the length of the longest run of ones that would be expected in a random sequence.

- **Binary Matrix Rank Test (BMRT)** is to check for linear dependence among fixed length substrings of the original sequence.

- **Discrete Fourier Transform (Spectral) Test (DFTT)** is to detect periodic features (i.e., repetitive patterns that are near each other) in the tested sequence that would indicate a deviation from the assumption of randomness.

- **Non-overlapping Template Matching Test (NOTMT)** is to detect generators that produce too many occurrences of a given non-periodic (aperiodic) pattern (m is the length in bits of each template which is the target string).

- **Overlapping Template Matching Test (OTMT)** is the number of occurrences of pre-specified target strings (m is the length in bits of the template - in this case, the length of the run of ones).

- **Maurer's "Universal Statistical" Test (MUST)** is to detect whether or not the sequence can be significantly compressed without loss of information (L is the length of each block, and Q is the number of blocks in the initialization sequence).

- **Linear Complexity Test (LCT)** is to determine whether or not the sequence is complex enough to be considered random (M is the length in bits of a block).

- **Serial Test (ST)** is to determine whether the number of occurrences of the 2^m m-bit (m is the length in bits of each block) overlapping patterns is approximately the same as would be expected for a random sequence.

- **Approximate Entropy Test (AET)** is to compare the frequency of overlapping blocks of two consecutive/adjacent lengths (m and m+1) against the expected result for a random sequence (m is the length of each block).

- **Cumulative Sums (Cusum) Test (CST)** is to determine whether the cumulative sum of the partial sequences occurring in the tested sequence is too large or too small relative to the expected behavior of that cumulative sum for random sequences.

- **Random Excursions Test (RET)** is to determine if the number of visits to a particular state within a cycle deviates from what one would expect for a random sequence.

- **Random Excursions Variant Test (REVT)** is to detect deviations from the expected number of visits to various states in the random walk.

DieHard battery of tests

The DieHard battery of tests was developed in 1996 by Prof. Georges Marsaglia from Florida State University, for testing randomness of sequences of numbers [103]. It has been the most sophisticated standard for over a decade. Because of the stringent requirements in the DieHard test suite, a generator passing the DieHard battery of tests can be considered good as a rule of thumb. It was supposed to give a better way of analysis in comparison to original FIPS statistical tests.

The DieHard battery of tests consists of 18 different independent statistical tests. Each test requires binary file of about 12 million bytes in order to run the full set of tests. As the NIST test suite, most of the tests in DieHard return a p-value, which should be uniform on $[0, 1)$ if the input file contains truly independent random bits. Those p-values are obtained by $p = F(X)$, where F is the assumed distribution of the sample random variable X (often normal). But that assumed F is just an asymptotic approximation, for which the fit will be worst in the tails. Thus occasional p-values near 0 or 1, such as 0.0012 or 0.9983 can occur. Unlike the NIST test suite, the test is considered to be successful when the p-value is in range $[0 + \alpha, 1 - \alpha]$, where α is the level of significance of the test.

For example, with a level of significance of 2.5%, $p - values$ are expected to be in $[0.025, 0.975]$. Note that if the p-value is not in this range, it means that the null hypothesis for randomness is rejected, even if the sequence is truly random (false negative). These tests are:

- **Birthday Spacings.** Choose random points on a large interval. The spacings between the points should be asymptotically Poisson distributed. The name is based on the birthday paradox.

- **Overlapping Permutations.** Analyze sequences of five consecutive random numbers. The 120 possible orderings should occur with statistically equal probability.

- **Ranks of matrices.** Select some number of bits from some number of random numbers to form a matrix over 0,1, then determine the rank of the matrix. Count the ranks.

- **Monkey Tests.** Treat sequences of some number of bits as "words." Count the overlapping words in a stream. The number of "words" that do not appear should follow a known distribution. The name is based on the infinite monkey theorem.

- **Count the 1's.** Count the 1 bits in each of either successive or chosen bytes. Convert the counts to "letters," and count the occurrences of five-letter "words."

- **Parking Lot Test.** Randomly place unit circles in a 100×100 square. If the circle overlaps an existing one, try again. After 12,000 tries, the number of successfully "parked" circles should follow a certain normal distribution.

- **Minimum Distance Test.** Randomly place 8,000 points in a $10,000 \times 10,000$ square, then find the minimum distance between the pairs. The square of this distance should be exponentially distributed with a certain mean.

- **Random Spheres Test.** Randomly choose 4,000 points in a cube of edge 1,000. Center a sphere on each point, whose radius is the minimum distance to another point. The smallest sphere's volume should be exponentially distributed with a certain mean.

- **The Squeeze Test.** Multiply 231 by random floats on $[0,1)$ until you reach 1. Repeat this 100,000 times. The number of floats needed to reach 1 should follow a certain distribution.

- **Overlapping Sums Test.** Generate a long sequence of random floats on $[0,1)$. Add sequences of 100 consecutive floats. The sums should be normally distributed with characteristic mean and sigma.

- **Runs Test.** Generate a long sequence of random floats on $[0,1)$. Count ascending and descending runs. The counts should follow a certain distribution.

- **The Craps Test.** Play 200,000 games of craps, counting the wins and the number of throws per game. Each count should follow a certain distribution.

Comparative test parameters

In this section, five well-known statistical tests [105] are used as comparison tools. They encompass frequency and autocorrelation tests. In what follows, $s = s^0, s^1, s^2, \ldots, s^{n-1}$ denotes a binary sequence of length n. The question is to determine whether this sequence possesses some specific characteristics that a truly random sequence would be likely to exhibit.

Frequency test (monobit test)

The purpose of this test is to check if the numbers of 0's and 1's are approximately equal in s, as it would be expected for a random sequence. Let n_0, n_1 denote these numbers. The statistic used here is

$$X_1 = \frac{(n_0 - n_1)^2}{n},$$

which approximately follows a χ^2 distribution with one degree of freedom when $n \geqslant 10^7$.

Serial test (2-bit test)

The purpose of this test is to determine if the number of occurrences of 00, 01, 10 and 11 as subsequences of s are approximately the same. Let n_{00}, n_{01}, n_{10}, and n_{11} denote the number of occurrences of $00, 01, 10$, and 11 respectively. Note that $n_{00} + n_{01} + n_{10} + n_{11} = n - 1$ since the subsequences are allowed to overlap. The statistic used here is:

$$X_2 = \frac{4}{n-1}(n_{00}^2 + n_{01}^2 + n_{10}^2 + n_{11}^2) - \frac{2}{n}(n_0^2 + n_1^2) + 1,$$

which approximately follows a χ^2 distribution with 2 degrees of freedom if $n \geqslant 21$.

Poker test

The poker test studies if each pattern of length m (without overlapping) appears the same number of times in s. Let $\lfloor \frac{n}{m} \rfloor \geqslant 5 \times 2^m$ and $k = \lfloor \frac{n}{m} \rfloor$. Divide the sequence s into k non-overlapping parts, each of length m. Let n_i be the number of occurrences of the ith type of sequence of length m, where $1 \leqslant i \leqslant 2^m$. The statistic used is

$$X_3 = \frac{2^m}{k} \left(\sum_{i=1}^{2^m} n_i^2 \right) - k,$$

which approximately follows a χ^2 distribution with $2^m - 1$ degrees of freedom. Note that the poker test is a generalization of the frequency test (setting $m = 1$ in the poker test yields the frequency test).

Runs test

The purpose of the runs test is to figure out whether the number of runs of various lengths in the sequence s is as expected, for a random sequence. A run is defined as a pattern of all zeros or all ones, a block is a run of ones, and a gap is a run of zeros. The expected number of gaps (or blocks) of length i in a random sequence of length n is $e_i = \frac{n-i+3}{2^{i+2}}$. Let k be equal to the largest integer i such that $e_i \geqslant 5$. Let B_i, G_i be the number of blocks and gaps of length i in s, for each $i \in [\![1, k]\!]$. The statistic used here will then be:

$$X_4 = \sum_{i=1}^{k} \frac{(B_i - e_i)^2}{e_i} + \sum_{i=1}^{k} \frac{(G_i - e_i)^2}{e_i},$$

which approximately follows a χ^2 distribution with $2k - 2$ degrees of freedom.

Autocorrelation test

The purpose of this test is to check for coincidences between the sequence s and (non-cyclic) shifted versions of it. Let d be a fixed integer, $1 \leqslant d \leqslant \lfloor n/2 \rfloor$. The $A(d) = \sum_{i=0}^{n-d-1} s_i \oplus s_{i+d}$ is the amount of bits not equal between the sequence and itself displaced by d bits. The statistic used is:

$$X_5 = \frac{2 \left(A(d) - \frac{n-d}{2} \right)}{\sqrt{n-d}},$$

which approximately follows a normal distribution $\mathcal{N}(0, 1)$ if $n - d \geqslant 10$. Since small values of $A(d)$ are as unexpected as large values, a two-sided test should be used.

TestU01 Statistical Test

TestU01 is the last battery of tests investigated in this book. This battery is extremely diverse in implementing classical tests, cryptographic tests, new tests proposed in the literature, and original tests. In fact, it encompasses most of the other test suites. In details, this TestU01 suite implements hundreds of tests and reports $p-values$ as previously. If a p-value is within $[0.001, 0.999]$, the associated test is a success. A p-value lying outside this boundary means that its test has failed.

The TestU01 package contains exactly seven batteries, which are listed below:

- **SmallCrush.** The first battery to check, with 15 $p-values$ reported. This is a fast collection of tests used to be sure that the basic requirements of randomness are satisfied. In case of success, this battery should be followed by Crush and BigCrush.

- **Crush.** This battery includes many difficult tests, like those described in [89]. It uses approximately 2^{35} random numbers and applies 96 statistical tests (it computes a total of 144 test statistics and $p-values$).

- **BigCrush.** This battery uses approximately 2^{38} random numbers and applies 106 tests (it computes 160 test statistics and $p-values$). A suite of very stringent statistical tests, and the most difficult battery to pass.

- **Rabbit.** This battery of tests reports 38 $p-values$.

- **Alphabit.** Alphabit and AlphabitFile have been designed primarily to test hardware random bits generators. 17 $p-values$ are reported.

- **Pseudo-DieHard.** This battery implements most of the tests contained in the popular battery DieHard or, in some cases, close approximations to them. It is not a very stringent battery. Indeed, there is no generator that can pass Crush and BigCrush batteries and fail Pseudo-DieHard, while the converse occurs for several defective generators. 126 $p-values$ are reported here.

- **FIPS_140_2.** As recalled previously, the NIST (National Institute of Standards and Technology) of the U.S. federal government has proposed a statistical test suite. It is used to evaluate the randomness of bitstreams produced by cryptographic random number generators. This battery reports 16 p-values.

Thus six predefined batteries of tests are available in TestU01. Three of them, namely the SmallCrush, Crush, and BigCrush batteries, are for sequences following a $\mathcal{U}(0,1)$ distribution (numbers supposed to be uniformly distributed in $[0,1]$), whereas the three others are for bit sequences. To test a RNG for general use, one could first apply the small and fast SmallCrush battery. In case of success, the more stringent Crush battery could then be applied, and if succeeded again, the yet more time-consuming BigCrush battery can finally be investigated.

Among other things, these batteries of tests include the classical tests described in Knuth [89]: the *run, poker, coupon collector, gap, max-of-t*, and *permutation* tests, for instance. Furthermore, these batteries embed collision and birthday spacings tests in 2, 3, 4, 7, and 8 dimensions, several close pairs tests in 2, 3, 5, 7, and 9 dimensions, and correlation tests too. Finally, some tests use the generated numbers as a sequence of "random" bits: random walk tests, linear complexity tests, a Lempel-Ziv compression test, several Hamming weights tests, matrix rank tests, run and correlation tests, among others.

As recalled previously, the Rabbit, Alphabit, and BlockAlphabit batteries are for binary sequences (*e.g.*, a cryptographic pseudorandom generator or a source of random bits produced by a physical device). They were originally designed to test a finite sequence contained in a binary file. When invoking one of these batteries, the number n_B of bits available for each test must be specified. When the bits are in a file, n_B must not exceed the number of bits in this file, and each test will reuse the same sequence of bits starting from the beginning of the file (so the tests are not independent). When the bits are produced by a generator, each test uses a different stream. In both cases, the parameters of each test are chosen automatically as a function of n_B .

The batteries Alphabit and Rabbit can be applied on a binary file considered as a source of random bits. They can also be applied on a programmed generator. Rabbit and Alphabit apply 38 and 17 different statistical tests respectively. Concerning the PseudoDieHard battery, it applies most of the tests in the well-known DieHard suite of Marsaglia detailed previously [103]. Last, the battery FIPS_140_2 implements the small suite of tests of the FIPS_140_2 standard from NIST. The batteries described in this module will write the results of each test (on standard output) with a standard level of details (assuming that the Boolean switches of module swrite have their default values), followed by a summary report of the suspect $p - values$ obtained from the specific tests included in the batteries. It is also possible to get only the summary report in the output, with no detailed output from the tests, by setting the Boolean switch swrite_Basic to FALSE.

Some of the tests compute more than one statistic (and p-value) using the same stream of random numbers, and these statistics are thus not independent. That is why the number of statistics in the summary reports is larger than the number of tests in the description of the batteries.

7.2.4.2   Test results for some well-known PRNGs

In this section, we first give the evaluation of ISAAC, XORshift, and the Logistic map generators on the batteries presented previously. We will then compare our proposed generators with these latter PRNGs.

In the experiments concerning the NIST tests suite, 100 sequences (s = 100) of 1,000,000 bits are generated and tested. If the value $\mathbb{P}_T$ of a given test is smaller than 0.0001, then the sequences are not random, and consequently the generator is claimed unsuitable. Table 7.7 shows $\mathbb{P}_T$ of sequences obtained from the Logistic map, XORshift, and ISAAC generators recalled in Section 7.2.3.1. If there are at least two statistical values in a test, this test is marked with an asterisk, and the average is computed to characterize the statistical values.

Table 7.8 gives the results derived from applying the DieHard battery

TABLE 7.7: NIST SP 800-22 test results ($\mathbb{P}_T$)

Test name	Logistic	XORshift	ISAAC
Frequency (Monobit) Test	0.53414	0.14532	0.67868
Frequency Test within a Block	0.00275	0.45593	0.10252
Runs Test	0.00001	0.21330	0.69931
Longest Run of Ones in a Block Test	0.08051	0.28966	0.43727
Binary Matrix Rank Test	0.67868	0.00000	0.89776
Discrete Fourier Transform (Spectral) Test	0.57490	0.00535	0.51412
Non-overlapping Template Matching Test*	0.28468	0.50365	0.55515
Overlapping Template Matching Test	0.10879	0.86769	0.63711
Universal Statistical Test	0.02054	0.27570	0.69931
Linear Complexity Test	0.79813	0.92407	0.03756
Serial Test* (m=10)	0.41542	0.75792	0.32681
Approximate Entropy Test (m=10)	0.02054	0.41902	0.30412
Cumulative Sums (Cusum) Test*	0.60617	0.81154	0.36786
Random Excursions Test*	0.53342	0.41923	0.50711
Random Excursions Variant Test*	0.28507	0.52833	0.40930
Success	15/15	14/15	15/15

TABLE 7.8: Results of DieHard battery of tests

No.	Test name	Logistic	XORshift	ISAAC
1	Overlapping Sum	Pass	Pass	Pass
2	Runs Up 1	Pass	Pass	Pass
	Runs Down 1	Pass	Pass	Pass
	Runs Up 2	Pass	Pass	Pass
	Runs Down 2	Pass	Pass	Pass
3	3D Spheres	Pass	Pass	Pass
4	Parking Lot	Pass	Pass	Pass
5	Birthday Spacing	Pass	Pass	Pass
6	Count the ones 1	Pass	Fail	Pass
7	Binary Rank 6×8	Pass	Pass	Pass
8	Binary Rank 31×31	Pass	Fail	Pass
9	Binary Rank 32×32	Pass	Fail	Pass
10	Count the ones 2	Pass	Pass	Pass
11	Bit Stream	Pass	Pass	Pass
12	Craps Wins	Pass	Pass	Pass
	Throws	Pass	Pass	Pass
13	Minimum Distance	Pass	Pass	Pass
14	Overlapping Perm.	Pass	Pass	Pass
15	Squeeze	Pass	Pass	Pass
16	OPSO	Pass	Pass	Pass
17	OQSO	Pass	Pass	Pass
18	DNA	Pass	Pass	Pass
	Number of tests passed	18	15	18

of tests to the three PRNGs first considered. We then show in Table 7.9 a comparison between the Logistic map, XORshift, and ISAAC generators through the comparative test parameters detailed in the previous section. Finally, Table 7.10 gives the results derived from applying the TestU01 battery of tests to the PRNGs considered here.

From the results of the TestU01, NIST, Comparative test parameters, and DieHard batteries of tests applied respectively to the Logistic map, XORshift, and ISAAC generators recalled in Section 7.2.3.1, we can deduce that the worst situation obviously appears when using either the Logistic map or the XORshift, and that ISAAC is able to pass the four batteries of tests.

TABLE 7.9: Comparative test parameters with a sequence having 10^7 bits

Method	Threshold values	Logistic	XORshift	ISAAC
Monobit	3.8415	0.1280	1.7053	0.1401
Serial	5.9915	0.1302	2.1466	0.1430
Poker	316.9194	240.2893	248.9318	236.8670
Runs	55.0027	26.5667	18.0087	34.1273
Autocorrelation	1.6449	0.0373	0.5099	-2.1712

TABLE 7.10: TestU01 Statistical Test

Test name	Battery	Nb of tests	Logistic	XORshift	ISAAC
Rabbit	32×10^9 bits	38	21	14	0
Alphabit	32×10^9 bits	17	16	9	0
Ps. DieHard	Standard	126	0	2	0
FIPS_140_2	Standard	16	0	0	0
SmallCrush	Standard	15	4	5	0
Crush	Standard	144	95	57	0
BigCrush	Standard	160	125	55	0
Failures			261	146	0

We can remark too that the Binary Matrix Rank Test (in NIST statistical test suite) is failed for the XORshift. This test focuses on the rank of disjoint sub-matrices of the entire sequence. Note that this latter also appears in the DieHard battery. Indeed, XORshift fails to pass three individual tests contained into the DieHard battery, namely the "Count the ones," "Binary Rank 31×31," and "Binary Rank 32×32" tests. We can thus conclude that, in the random numbers obtained with XORshift, only the least significant bits seem to be really independent. This fact can explain the poor behavior of this PRNG in the aforementioned basic tests that evaluate the independence of real numbers.

Seven tests have been failed with a p-value practically equal to 0 or 1 for both the Logistic map and XORshift generators, when considering the TestU01 Statistical Test. It is clear that, with such results, the null hypothesis H_0 claiming that, for each integer $t > 0$, the vector $(u_0, ..., u_{t-1})$ is uniformly distributed over the t-dimensional unit cube $[0,1]^t$, must be rejected for these two bit streams.

7.2.4.3 Test results and comparative analysis for the proposed Old CI

In a sound theoretical basis, the Old CI PRNG based on discrete chaotic iterations of Chapter 5 is a composite generator that combines the features of two PRNGs. The first generator constitutes the initial condition of the chaotic dynamical system. The second one randomly chooses which outputs of the chaotic system must be returned. The intention of this combination is to accumulate the effects of chaotic and random behaviors, in order to improve the statistical and security properties relative to each generator taken alone.

This PRNG based on discrete chaotic iterations may utilize any reasonable RNG as inputs. For demonstration purposes, Logistic map, XORshift and ISAAC are adopted here. As the Old CI PRNG depends on various parameters, the first task is to determine good values for these latter.

How to find good parameters

To exhibit the correlation between the parameter k such that $\mathcal{M} = \{k, k+1\}$ and the success rate, as discussed in Section 7.2.3.2, we have used CI(ISAAC, XORshift) PRNG as a concrete example. In Figure 7.10 is plotted the SmallCrush test on sequences generated by CI(ISAAC, XORshift). SmallCrush is succeeded when the total number of accepted tests is 15. In these figures, the ordinates are the number of successful passes a sequence goes through. Thus, a qualified sequence must have most of the time a passing value equal to 15, with possible occasional failures (even a true random sequence can occasionally fail these tests). It can be seen in Figure 7.10, and it has been obtained too in other simulations we have realized, that when $k > 3N$, the sequences tend to pass the SmallCrush test.

To compare the sequences generated with different parameters N in a more quantitative manner, we have set $k = 3N+1$, and the same analysis for the four CI(X,Y) generators through TestU01 has been repeated. Figure 7.11 shows the number of passing sequences generated by the CI(X,Y) PRNGs proposed in this book, with the same parameters and initial values. Obviously, $N = 4$ gives the best results. For this reason, we will consider in the following sections that $N = 4$ and $k = 12$.

Results returned by the batteries of tests

In the experiments that concern the NIST battery, 100 sequences ($s = 100$) of 1,000,000 bits are generated and tested. If the value $\mathbb{P}_T$ of one test is smaller than 0.0001, the sequences are not random. Table 7.11 shows $\mathbb{P}_T$ of the sequences obtained with the Old CI PRNG using different schemes. As previously, if there are at least two statistical values in a test, this latter is marked with an asterisk, and the average value is computed to characterize the statistics.

Table 7.12 gives the results derived from applying the DieHard battery of tests to the PRNGs considered here.   Additionally, we have given in Table 7.13 the comparative test parameters between Old CI(Logistic, Logistic), Old CI(XORshift, XORshift), Old CI(ISAAC, XORshift), and Old

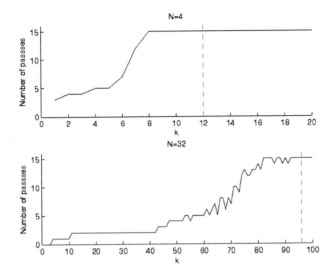

FIGURE 7.10: SmallCrush for CI(ISAAC,XORshift)

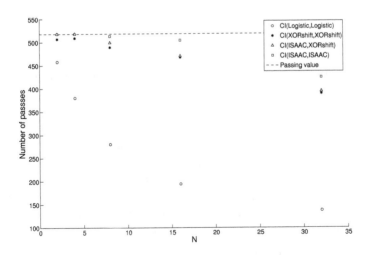

FIGURE 7.11: TestU01 results

TABLE 7.11: NIST SP 800-22 test results ($\mathbb{P}_T$) for Old CI algorithms (N = 4)

Test name	Old CI			
	Logistic + Logistic	XORshift + XORshift	ISAAC + XORshift	ISAAC + ISAAC
Frequency (Monobit) Test	0.85138	0.59554	0.40119	0.33453
Frequency Test within a Block	0.38382	0.55442	0.89776	0.71974
Runs Test	0.31908	0.45593	0.31908	0.38382
Longest Run of Ones in a Block Test	0.13728	0.01671	0.08558	0.67868
Binary Matrix Rank Test	0.69931	0.61630	0.47498	0.79813
Discrete Fourier Transform (Spectral)	0.12962	0.00019	0.77918	0.67868
Non-overlapping Template Matching*	0.48473	0.53225	0.53568	0.51258
Overlapping Template Matching Test	0.47498	0.33453	0.36691	0.07571
Universal Statistical Test	0.09657	0.03292	0.26224	0.85138
Linear Complexity Test	0.41902	0.40119	0.61715	0.21330
Serial Test* (m=10)	0.53427	0.01339	0.33453	0.76102
Approximate Entropy Test (m=10)	0.99146	0.13728	0.53414	0.22482
Cumulative Sums (Cusum) Test*	0.75530	0.04646	0.31915	0.47658
Random Excursions Test*	0.65406	0.50362	0.50804	0.46305
Random Excursions Variant Test*	0.55388	0.34777	0.48400	0.54863
Success	15/15	15/15	15/15	15/15

CI(ISAAC, ISAAC). Finally, Table 7.14 gives the results derived from applying the TestU01 battery of tests to the PRNGs considered in this section.

The results of comparative test parameters confirm that the proposed CI PRNGs are all able to pass these tests. Statistical results of comparative test parameters for both CI(XORshift, XORshift) and CI(ISAAC, XORshift) are better for most of the parameters, leading to the conclusion that these generators are more random than the former ones. This improvement clearly appears in the TestU01 results: XORshift alone fails 142 of these tests, whereas CI(XORshift, XORshift) only fails 9 out of 518.

In other words, in addition of having chaotic properties, the Old CI PRNG based on discrete chaotic iterations can pass more performed tests than its individual components taken alone.

We now give an evaluation of the New CI version of our proposed generator.

TABLE 7.12: Results of DieHard battery of tests for Old CI algorithms (N = 4)

| No. | Test name | Old CI | | | |
		Logistic + Logistic	XORshift + XORshift	ISAAC + XORshift	ISAAC + ISAAC
1	Overlapping Sum	Pass	Pass	Pass	Pass
2	Runs Up 1	Pass	Pass	Pass	Pass
	Runs Down 1	Pass	Pass	Pass	Pass
	Runs Up 2	Pass	Pass	Pass	Pass
	Runs Down 2	Pass	Pass	Pass	Pass
3	3D Spheres	Pass	Pass	Pass	Pass
4	Parking Lot	Pass	Pass	Pass	Pass
5	Birthday Spacing	Pass	Pass	Pass	Pass
6	Count the ones 1	Pass	Pass	Pass	Pass
7	Binary Rank 6 × 8	Pass	Pass	Pass	Pass
8	Binary Rank 31 × 31	Pass	Pass	Pass	Pass
9	Binary Rank 32 × 32	Pass	Pass	Pass	Pass
10	Count the ones 2	Pass	Pass	Pass	Pass
11	Bit Stream	Pass	Pass	Pass	Pass
12	Craps Wins	Pass	Pass	Pass	Pass
	Throws	Pass	Pass	Pass	Pass
13	Minimum Distance	Pass	Pass	Pass	Pass
14	Overlapping Perm.	Pass	Pass	Pass	Pass
15	Squeeze	Pass	Pass	Pass	Pass
16	OPSO	Pass	Pass	Pass	Pass
17	OQSO	Pass	Pass	Pass	Pass
18	DNA	Pass	Pass	Pass	Pass
	Number of tests passed	18	18	18	18

7.2.4.4 Test results and comparative analysis for the New CI PRNG

Let us recall that this New CI PRNG is a composite generator, which combines the features of two other PRNGs: the first generator constitutes

TABLE 7.13: Comparative test parameters for Old CI(X,Y) with a 10^7 bits sequence (N = 4)

		Old CI			
Method	Threshold values	Logistic + Logistic	XORshift + XORshift	ISAAC + XORshift	ISAAC + ISAAC
Monobit	3.8415	1.0368	3.5689	0.0569	0.6641
Serial	5.9915	1.1758	3.5765	0.9828	0.6506
Poker	316.9194	269.0607	222.3683	243.8415	262.6440
Runs	55.0027	36.5479	28.4237	29.3195	30.3116
Autocor.	1.6449	0.4054	0.3403	0.6141	0.9455

TABLE 7.14: TestU01 Statistical Test for Old CI algorithms (N = 4)

			Old CI			
Test name			Logistic + Logistic	XORshift + XORshift	ISAAC + XORshift	ISAAC + ISAAC
Rabbit	32×10^9 b.	38	7	2	0	0
Alphabit	32×10^9 b.	17	3	0	0	0
Ps. DieHard	Standard	126	0	0	0	0
FIPS_140_2	Standard	16	0	0	0	0
SmallCrush	Standard	15	2	0	0	0
Crush	Standard	144	47	4	0	0
BigCrush	Standard	160	79	3	0	0
Failures		518	138	9	0	0

the initial condition of the chaotic dynamical system, whereas the second generator randomly chooses, after decimation, which outputs of the chaotic system must be returned. The intention of this combination is the same as for the Old CI PRNG: to accumulate the effects of chaotic and random behaviors, to improve the statistical and security properties relative to each generator taken alone.

This New CI PRNG based on discrete chaotic iterations may utilize any

TABLE 7.15: SP 800-22 test results ($\mathbb{P}_T$) for new CI(XORshift, XORshift) (N = 32, and * means that two statistical values are averaged)

Method	New CI		
	$m^n = y^n \ mod \ N$	no mark	$g_2()$
Frequency (Monobit) Test	0.00040	0.08556	0.41943
Frequency Test within a Block	0	0	0.67862
Runs Test	0.28966	0.55448	0.33452
Longest Run of Ones in a Block	0.01096	0.43723	0.88313
Binary Matrix Rank Test	0	0.65794	0.75972
Discrete Fourier Transform	0	0	0.00085
Non-overlapping Template Match.*	0.02007	0.37333	0.51879
Overlapping Template Matching	0	0	0.24924
Maurer's "Universal Statistical"	0.69936	0.96424	0.12963
Linear Complexity Test	0.36699	0.92423	0.35045
Serial Test* (m=10)	0	0.28185	0.25496
Approximate Entropy Test (m=10)	0	0.38381	0.75971
Cumulative Sums (Cusum) Test*	0	0	0.34245
Random Excursions Test*	0.46769	0.34788	0.18977
Random Excursions Variant Test*	0.28779	0.46505	0.26563
Success	8/15	11/15	15/15

reasonable RNG as inputs. However, for demonstration purposes, we have adopted XORshift and ISAAC here.

Results produced by the batteries of tests

The New CI generator has been tested like the other PRNGs studied in this chapter. Tables 7.15 and 7.16 show $\mathbb{P}_T$ of NIST battery, for sequences obtained with the New CI generator. We can conclude from Table 7.15 that the worst situations are obtained with the New CI ($m^n = y^n \ mod \ N$) and New CI (no mark) generators. Contrarily, New CI ($m^n = g_2(y^n)$) and New CI ($m^n = g_1(y^n)$) have successfully passed the NIST statistical test suite (see Table 7.16).

Table 7.17 shows the results derived from applying the DieHard battery of tests to the PRNGs considered in this section. Obviously, these generators have no problem successfully passing the DieHard battery.

We then show in Table 7.18 a comparison between New CI(XORshift, XORshift), New CI(ISAAC, XORshift), and New CI(ISAAC, ISAAC), when considering the comparative test parameters. No bias can be reported here.

TABLE 7.16: NIST SP 800-22 test results ($\mathbb{P}_T$) for new CI algorithms

Test name	New CI XORshift + XORshift	ISAAC + XORshift	ISAAC + ISAAC
Frequency (Monobit) Test	0.47498	0.88317	0.83430
Frequency Test within a Block	0.89776	0.40119	0.33453
Runs Test	0.81653	0.31908	0.00576
Longest Run of Ones in a Block Test	0.79813	0.06688	0.47498
Binary Matrix Rank Test	0.26224	0.88317	0.69931
Discrete Fourier Transform (Spectral) Test	0.00716	0.33453	0.59559
Non-overlapping Template Matching Test*	0.44991	0.46467	0.51446
Overlapping Template Matching Test	0.51412	0.69931	0.88317
Universal Statistical Test	0.67868	0.24928	0.06282
Linear Complexity Test	0.65793	0.65793	0.94630
Serial Test* (m=10)	0.42534	0.90619	0.44137
Approximate Entropy Test (m=10)	0.63719	0.22482	0.13728
Cumulative Sums (Cusum) Test*	0.27968	0.84065	0.14139
Random Excursions Test*	0.28740	0.30075	0.34625
Random Excursions Variant Test*	0.48668	0.34294	0.55048
Success	15/15	15/15	15/15

Finally, Table 7.19 gives the results derived from applying the TestU01 battery of tests to the PRNGs considered here. These results are good enough to consider the generators as random. They all can pass this very stringent battery.

The results of the TestU01, NIST, Comparative test parameters, and DieHard batteries of tests confirm that the proposed New CI PRNGs are all able to pass these tests. Results are quite better than for the Old CI PRNG. The improvement clearly appears in the TestU01 results. We recall that the XORshift generator taken alone fails 142 of these tests, whereas Old CI(XORshift, XORshift) fails 9 out of 518. We can see that the New CI(XORshift, XORshift) is able to pass all the tests, which is obviously what we intended to show. Detailed comparative analysis of this improvement will now be presented in the next section.

TABLE 7.17: Results of DieHard battery of tests for new CI algorithms (N = 32)

No.	Test name	New CI			
		Logistic + Logistic	XORshift + XORshift	ISAAC + XORshift	ISAAC + ISAAC
1	Overlapping Sum	Pass	Pass	Pass	Pass
2	Runs Up 1	Pass	Pass	Pass	Pass
	Runs Down 1	Pass	Pass	Pass	Pass
	Runs Up 2	Pass	Pass	Pass	Pass
	Runs Down 2	Pass	Pass	Pass	Pass
3	3D Spheres	Pass	Pass	Pass	Pass
4	Parking Lot	Pass	Pass	Pass	Pass
5	Birthday Spacing	Pass	Pass	Pass	Pass
6	Count the ones 1	Pass	Pass	Pass	Pass
7	Binary Rank 6 × 8	Pass	Pass	Pass	Pass
8	Binary Rank 31 × 31	Pass	Pass	Pass	Pass
9	Binary Rank 32 × 32	Pass	Pass	Pass	Pass
10	Count the ones 2	Pass	Pass	Pass	Pass
11	Bit Stream	Pass	Pass	Pass	Pass
12	Craps Wins	Pass	Pass	Pass	Pass
	Throws	Pass	Pass	Pass	Pass
13	Minimum Distance	Pass	Pass	Pass	Pass
14	Overlapping Perm.	Pass	Pass	Pass	Pass
15	Squeeze	Pass	Pass	Pass	Pass
16	OPSO	Pass	Pass	Pass	Pass
17	OQSO	Pass	Pass	Pass	Pass
18	DNA	Pass	Pass	Pass	Pass
	Number of tests passed	18	18	18	18

7.2.4.5 Further investigations of two chaotic iteration schemes based on the XORshift generator

We will now compare, in this section, the simple XORshift generator with the two improved versions we proposed, namely the Old CI(XORshift, XORshift) and the New CI(XORshift, XORshift) PRNGs.

TABLE 7.18: Comparative test parameters for new CI(X,Y) with a 10^7 bits sequence (N = 32)

Method	Threshold values	New CI		
		XORshift + XORshift	ISAAC + XORshift	ISAAC + ISAAC
Monobit	3.8415	3.5689	0.9036	0.5788
Serial	5.9915	3.5765	1.1229	0.7378
Poker	316.9194	123.6831	173.8604	179.8609
Runs	55.0027	28.4237	40.4606	29.4057
Autocor.	1.6449	0.3403	0.1245	-2.0276

TABLE 7.19: TestU01 Statistical Battery for New CI (N = 32)

Test name			New CI		
			Logistic + Logistic	ISAAC + XORshift	ISAAC + ISAAC
Rabbit	32×10^9 bits	38	0	0	0
Alphabit	32×10^9 bits	17	0	0	0
Ps. DieHard	Standard	126	0	0	0
FIPS_140_2	Standard	16	0	0	0
SmallCrush	Standard	15	0	0	0
Crush	Standard	144	0	0	0
BigCrush	Standard	160	0	0	0
Number of failures			0	0	0

Let us recall the experimental protocol for the NIST battery: 100 sequences (s = 100) of 1,000,000 bits are generated and tested. If the value $\mathbb{P}_T$ of any test is smaller than 0.0001, the sequences are considered to not be good enough and the generator is unsuitable. Table 7.20 gives $\mathbb{P}_T$ of the sequences generated by the different schemes considered here. We can show in Table 7.20 that the XORshift taken alone has failed 1 test, whereas both the Old and New CI(XORshift, XORshift) generators have successfully passed the NIST statis-

TABLE 7.20: NIST SP 800-22 test results (N = 32, * means that two statistical values are averaged)

Method	XORshift	Old CI	New CI
Frequency (Monobit) Test	0.145326	0.595549	0.474986
Frequency Test within a Block	0.455937	0.554420	0.897763
Runs Test	0.213309	0.455937	0.816537
Longest Run of Ones in a Block Test	0.289667	0.016717	0.798139
Binary Matrix Rank Test	0.000000	0.616305	0.262249
Discrete Fourier Transform (Spectral) Test	0.005358	0.000190	0.007160
Non-overlapping Template Matching Test*	0.503652	0.532252	0.449916
Overlapping Template Matching Test	0.867692	0.334538	0.514124
Universal Statistical Test	0.275709	0.032923	0.678686
Linear Complexity Test	0.924076	0.401199	0.657933
Serial Test* (m=10)	0.757925	0.013396	0.425346
Approximate Entropy Test (m=10)	0.419021	0.137282	0.637119
Cumulative Sums (Cusum) Test*	0.8115445	0.046464	0.279680
Random Excursions Test*	0.4192395	0.503622	0.287409
Random Excursions Variant Test*	0.5283333	0.347772	0.486686
Success	14/15	15/15	15/15

tical test suite. This result shows the good behavior of the proposed PRNGs in the battery considered here.

Table 7.21 gives the results derived from applying the DieHard battery of tests to the PRNGs considered in this section. As stated previously, these results show that, in the random numbers obtained with the XORshift generator, only the least significant bits seem to be independent, which explains the poor behavior of this PRNG in this battery of tests. Contrarily, the generators based on discrete chaotic iterations (the Old CI and New CI PRNGs) can pass all the DieHard battery of tests. This proves that the randomness of the given generator has been improved by chaotic iterations.

We now consider the comparative test parameters in Table 7.22, between the New CI(XORshift, XORshift) generator, its old version denoted Old CI(XORshift, XORshift), and a PRNG based on a simple XORshift. Time (in seconds) is related to the duration needed by each algorithm to generate a 2×10^5 bits long sequence. The tests have been conducted using the same computer and compiler with the same optimization settings for both algorithms, in order to make them as fair as possible. The results confirm that

TABLE 7.21: Results when considering the DieHard battery (N = 32)

No.	Test name	Generators		
		XORshift	Old CI	New CI
1	Overlapping Sum	Pass	Pass	Pass
2	Runs Up 1	Pass	Pass	Pass
	Runs Down 1	Pass	Pass	Pass
	Runs Up 2	Pass	Pass	Pass
	Runs Down 2	Pass	Pass	Pass
3	3D Spheres	Pass	Pass	Pass
4	Parking Lot	Pass	Pass	Pass
5	Birthday Spacing	Pass	Pass	Pass
6	Count the ones 1	Fail	Pass	Pass
7	Binary Rank 6 × 8	Pass	Pass	Pass
8	Binary Rank 31 × 31	Fail	Pass	Pass
9	Binary Rank 32 × 32	Fail	Pass	Pass
10	Count the ones 2	Pass	Pass	Pass
11	Bit Stream	Pass	Pass	Pass
12	Craps Wins	Pass	Pass	Pass
	Throws	Pass	Pass	Pass
13	Minimum Distance	Pass	Pass	Pass
14	Overlapping Perm.	Pass	Pass	Pass
15	Squeeze	Pass	Pass	Pass
16	OPSO	Pass	Pass	Pass
17	OQSO	Pass	Pass	Pass
18	DNA	Pass	Pass	Pass
	Number of tests passed	15	18	18

the proposed generator is a lot faster than the old one, while the statistical results are better for most of the parameters, leading to the conclusion that the new PRNG is better than the old one.

As a comparison of the overall stability of these PRNGs, similar tests have been computed for different sequence lengths (see Figures 7.12 to 7.16). For the monobit test comparison (Figure 7.12), XORshift and New CI(XORshift, XORshift) PRNGs present the same issue: the beginning values are a little high. However, for our new generator, the values are stable in a low level

TABLE 7.22: Comparison with Old CI(XORshift,XORshift) for a 2×10^5 bits sequence ($N = 32$)

Method	The threshold values	XORshift	Old CI	New CI
Monobit	3.8415	1.7053	2.7689	0.3328
Serial	5.9915	2.1466	2.8765	0.7441
Poker	316.9194	248.9318	222.3683	262.8173
Runs	55.0027	18.0087	21.9272	16.7877
Autocorrelation	1.6449	0.5009	0.0173	0.0805
Time	Second	0.096s	0.3625s	0.197s

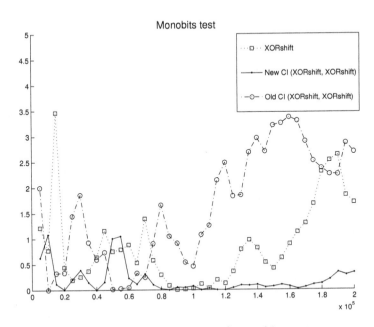

FIGURE 7.12: Comparison of monobits tests

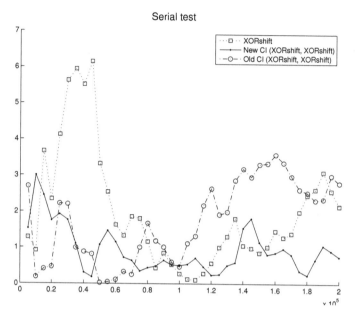

FIGURE 7.13: Comparison of serial tests

which never exceeds 1.2. Indeed, the new generator distributes very randomly the zeros and ones, whatever the length of the desired sequence. It can also be remarked that the old generator presents the worst performance, but the values are still within the standard boundary.

Figure 7.13 shows the serial test comparison. The new generator outperforms this test, but the score of the old generator is not bad either; their occurrences of 00, 01, 10, and 11 are very close to each other.

The poker test comparison with $m = 8$ is shown in Figure 7.14. XORshift is the most stable generator in all of these tests, but it is not better than the Old CI(XORshift, XORshift) PRNG. Our new generator presents a trend, with a maximum in the neighborhood of 1.7×10^5. These scores are not so good, even though the new generator has a better behavior than the old one and XORshift. Indeed, the value of m and the length of the sequences should be enlarged to be certain that the chaotic iterations express totally their complex behavior. In that situation, the performances of our generators in the poker test can be improved.

The graph of the new generator is the most stable one during the runs test comparison (Figure 7.15). Moreover, this trend is reinforced when the lengths of the tested sequences are increased.

The comparison of autocorrelation tests is presented in Figure 7.16. The new generator clearly dominates these tests, whereas the score of the old generator is also not bad. This difference between two generators based on

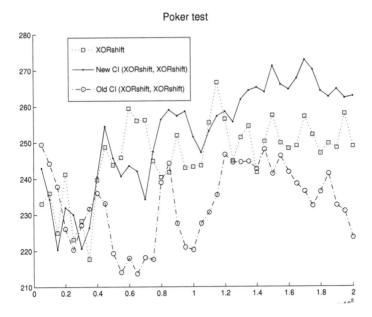

FIGURE 7.14: Comparison of poker tests

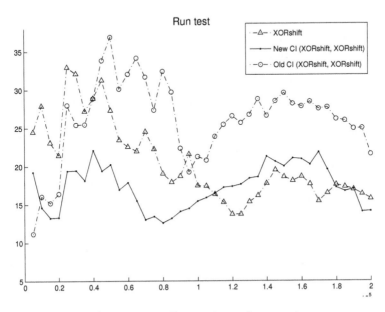

FIGURE 7.15: Comparison of run tests

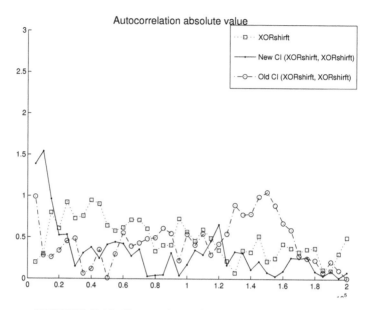

FIGURE 7.16: Comparison of autocorrelation tests

chaotic iterations can be explained by the fact that the improvements realized to define the new generator lead to a more random output.

To sum up, we can claim that the new generator, which is faster than its former version, outperforms all of the other generators in these statistical tests, especially when producing long output sequences.

A flexible output

We assume that the initial state X is given as an array of N-bit integers. Thus, the output size can be flexibly chosen by changing N. Moreover, due to the fact that the CI process is a simple bitwise change, the speed of output integers and binary numbers is almost the same.

In the following, we will investigate the randomness level of the generators for various N. For both CI generators, various N can pass all the NIST and DieHard tests. Table 7.23 gives the results derived from applying the TestU01 battery of tests to the PRNGs considered in this section. As can be observed, the effective range of N for the New CI is bigger than the one for the Old CI according to the battery TestU01.

It can be concluded from this experimental study that the new way of obtaining a PRNG by combining two XORshifts has better statistical properties than the old one, which is better than the individual XORshift taken alone (this latter fails 146 tests in TestU01).

TABLE 7.23: TestU01 Statistical Test

CI PRNG	Battery	N=2	N=4	N=8	N=16	N=32
	Rabbit	2	2	2	2	3
	Alphabit	0	0	0	2	2
	Pseudo DieHard	0	0	0	0	0
Old CI	FIPS_140_2	0	0	0	0	0
(XORshift, XORshift)	SmallCrush	0	0	0	1	0
	Crush	4	4	9	16	46
	BigCrush	5	3	18	30	78
	Number of failures	11	9	29	51	129
	Rabbit	0	0	0	0	0
	Alphabit	4	0	0	0	0
	Pseudo DieHard	8	2	0	0	0
New CI	FIPS_140_2	2	0	0	0	0
(XORshift, XORshift)	SmallCrush	0	0	0	0	0
	Crush	0	0	0	0	0
	BigCrush	0	0	0	0	0
	Number of failures	14	2	0	0	0

7.3   Hash Functions

In this section, another concrete example of a chaotic program is given in the computer science security field.

7.3.1   Introduction

The use of chaotic maps to generate hash algorithms has seen several developments in recent years. In [64] for example, a digital signature algorithm based on an elliptic curve and chaotic mapping is proposed to strengthen the security of an elliptic curve digital signature algorithm. Other examples of the generation of a hash function using chaotic maps can be found in, *e.g.*, [110, 137, 140]. Neural networks that have learned a continuous chaotic map have been proposed too in recent years [93], to achieve hash function requirements.

Note that using any chaotic map does not guarantee that the resulting hash function would behave chaotically too. To the best of our knowledge, this point is not discussed in the referenced papers, however, it should be considered as important. In this section we define a way to construct hash functions based on chaotic iterations. As a consequence of the theory presented before, the generated hash functions satisfy various topological chaos properties.

7.3.2   A chaotic hash function

In this section, we give a way to obtain a digest of a digital medium described by a binary sequence. It is based on chaotic iterations and satisfies various topological chaos properties. The hash value will be the last state of some chaotic iterations; the initial state X_0, finite strategy S, and iterate function must then be defined.

The initial condition $X_0 = (S, E)$ is composed by a $N = 256$ bit sequence E and a chaotic strategy S. In the following section, we describe in detail how to obtain this initial condition from the original medium.

7.3.2.1   How to obtain E

The first step of our algorithm is to transform the message in a normalized 256 bit sequence E. To illustrate this step inspired by SHA-1, we state that our original text is: "*The original text.*" Each character of this string is replaced by its ASCII code (on 7 bits). Then, we add a 1 at the end of this string.

```
10101001 10100011 00101010 00001101 11111100 10110100
11100111 11010011 10111011 00001110 11000100 00011101
00110010 11111000 11101001
```

So, the binary value (1111000) of the length of this string (120) is added, with another 1:

```
10101001 10100011 00101010 00001101 11111100 10110100
11100111 11010011 10111011 00001110 11000100 00011101
00110010 11111000 11101001 11110001
```

This string is inverted (the last bit is now the first one) and the two new substrings are concatenated. This gives:

```
10101001 10100011 00101010 00001101 11111100 10110100
11100111 11010011 10111011 00001110 11000100 00011101
00110010 11111000 11101001 11110001 00011111 00101110
00111110 10011001 01110000 01000110 11100001 10111011
10010111 11001110 01011010 01111111 01100000 10101001
10001011 0010101
```

So, we obtain a multiple of 512, by duplicating this string enough and truncating at the next multiple of 512. This string in which the whole original text is contained, is denoted by D.

Finally, we split the new string into blocks of 256 bits and apply the exclusive-or function, obtaining a 256 bits sequence in a manner inspired by the SHA-X algorithms.

```
11111010 11100101 01111110 00010110 00000101 11011101
00101000 01110100 11001101 00010011 01001100 00100111
01010111 00001001 00111010 00010011 00100001 01110010
01000011 10101011 10010000 11001011 00100010 11001100
10111000 01010010 11101110 10000001 10100001 11111010
10011101 01111101
```

In the context of Subsection 7.3.2, $N = 256$, and E is the above obtained sequence of 256 bits: the given message has been compressed into a 256 binary string.

We now have the definitive length of our digest. Note that a lot of texts have the same normalized string. This is not a problem because the strategy we will build depends on the whole text too, in such a way that two different texts lead to two different strategies. Let us now build the strategy S.

7.3.2.2   How to choose S

To obtain the strategy S, an intermediate sequence (u^n) is constructed from D as follows:

- D is split into blocks of 8 bits. Then u^n is the decimal value of the nth block.

- A circular rotation of one bit to the left is applied to D (the first bit of D is put on the end of D). Then the new string is split into blocks of 8 bits another time. The decimal values of those blocks are added to (u^n).

- This operation is repeated again 6 times.

It is now possible to build the strategy S:

$$S^0 = u^0, \quad S^n = (u^n + 2 \times S^{n-1} + n) \ (mod \ 256).$$

S will be highly dependent on the changes of the original text, as we have recalled that $\theta \longmapsto 2\theta \ (mod \ 1)$ is chaotic as defined by Devaney's theory.

7.3.2.3   How to construct the digest

To construct the digest, chaotic iterations are done with initial state X^0,

$$f : \quad \begin{matrix} [\![1, 256]\!] \\ (E_1, \ldots, E_{256}) \end{matrix} \quad \begin{matrix} \longrightarrow \\ \longmapsto \end{matrix} \quad \begin{matrix} [\![1, 256]\!] \\ (\overline{E_1}, \ldots, \overline{E_{256}}), \end{matrix}$$

as iterate function, and S for the chaotic strategy.

The result of these iterations is a 256 bits vector. Its components are taken 4 bits at a time and translated into hexadecimal numbers, to obtain the hash value:

63A88CB6AF0B18E3BE828F9BDA4596A6A13DFE38440AB9557DA1C0C6B1EDBDBD

To compare, if instead of using the text "*The original text*" we took "*the original text,*" the hash function returns:

33E0DFB5BB1D88C924D2AF80B14FF5A7B1A3DEF9D0E831194BD814C8A3B948B3

The generation of hash value is done with the vectorial Boolean negation f_0 defined in Eq. (5.2). Nevertheless, the procedure remains general and can be applied with any function f such that G_f is chaotic. In the following subsection, a complete example of the procedure is given.

(a) Original image.   (b) Modified image.

FIGURE 7.17: Hash of some black and white images.

7.3.3   Application example

Consider two black and white images of size 64×64 in Fig. 7.17, in which the pixel in position (40,40) has been changed. In this case, the hash function returns:

34A5C1B3DFFCC8902F7B248C3ABEFE2C9C9538E5104D117B399C999F74CF1CAD

for the Figure 7.17a and

5E67725CAA6B7B7434BE57F5F30F2D3D57056FA960B69052453CBC62D9267896

for the Figure 7.17b.

Consider two 256 graylevel images of Lena (256×256 pixels) in Figure 7.18, in which the grayscale level of the pixel in position (50,50) has been transformed from 93 (Figure 7.18a) to 94 (Figure 7.18b). In this case, the

(a) Original Lena.   (b) Modified Lena.

FIGURE 7.18: Hash of some grayscale level images.

hash function returns:

FA9F51EFA97808CE6BFF5F9F662DCD738C25101FE9F7F427CD4E2B8D40331B89 for the left Lena and

BABF2CE1455CA28F7BA20F52DFBD24B76042DC572FCCA4351D264ACF4C2E108B

for the right Lena.

These examples give an illustration of the avalanche effect obtained by this algorithm.

7.3.4   A chaotic neural network as hash function

A hash function can be achieved in two stages: the compression of the message (mapping a binary sequence of any length $n \in \mathbb{N}$ into a message of a fixed length belonging into $\mathbb{B}^N$, for a given fixed length $N \in \mathbb{N}$) and the hash of the compressed message [93]. As several compression functions have yet been proposed to achieve the first stage, we will only focus on the second stage and we will explain how to build a neural network that realizes it. This neural network that hashes compressed messages will behave chaotically, as it is defined by Devaney's theory. The reader is referred to [10, 19, 25, 77] for further details about the proposed hash function and its relationship with chaotic neural networks.

Let us first explain how it is possible to build a neural network that behaves chaotically. Consider $f : \mathbb{B}^N \longrightarrow \mathbb{N}^N$ and a MLP that recognizes F_f. That means, for all $(k, x) \in [\![1; N]\!] \times \mathbb{B}^N$, the response of the output layer to the input (k, x) is $F_f(k, x)$. We thus connect the output layer to the input one as it is depicted in Figure 7.19, leading to a global recurrent artificial neural network (ANN) working as follows [33].

- At the initialization stage, the ANN receives a Boolean vector $x^0 \in \mathbb{B}^N$ as input state, and $S^0 \in [\![1; N]\!]$ in its input integer channel $i()$. Thus, $x^1 = F_f(S^0, x^0) \in \mathbb{B}^N$ is computed by the neural network.

- This state x^1 is published as an output. Additionally, x^1 is sent back to the input layer, to act as Boolean state in the next iteration.

- At iteration number n, the recurrent neural network receives the state $x^n \in \mathbb{B}^N$ from its output layer and $i(S^n) \in [\![1; N]\!]$ from its input integer channel $i()$. It can thus calculate $x^{n+1} = F_f(i(S^n), x^n) \in \mathbb{B}^N$, which will be the new output of the network.

Obviously, this particular MLP produces exactly the same values as CIs with update function f. That is, such MLPs are equivalent to CIs with f as update function. However, the compression stage of the hash function presented in the previous section can be resumed to make chaotic iterations over the compressed message. As chaotic iterations can be obtained with a neural network, we can thus realize this stage with a (chaotic) neural network. Finally, it is important to remark that the proposed hash function can be

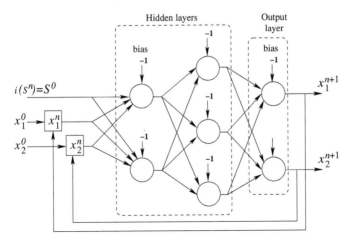

FIGURE 7.19: Example of global recurrent neural network modeling function F_f such that $x^{n+1} = \left(x_1^{n+1}, x_2^{n+1} \right) = F_f \left(i(S^n), (x_1^n, x_2^n) \right)$

implemented in a global neural network, as various compression neural networks can be found in the literature [59, 99, 122]: we just have to replace our compression stage, inspired by SHA-X, with a compression ANN.

Chapter 8

Wireless Sensor Networks

Wireless sensor networks (WSN) are sensors on batteries that are able to communicate with each other using a wireless communication. They are often reputed as promising tools in health care (body area networks, home support of seniors or dependent persons), monitoring (video surveillance, battlefields), and natural disaster management [5]. As they can be deployed in hostile environments, security is often an important element that must be taken into account [6]. However, due to their restricted resources and batteries, usual security solutions like AES or RSA cryptosystems are not relevant to provide security in communications, among other things due to their computation costs. Authors of this book have investigated the discrete chaotic dynamical system approach in [20–24, 29], providing by doing so new original and relevant solutions to security issues of WSNs like video surveillance (Section 8.1), secure scheduling (Sect. 8.1.5.1), coverage (Sect. 8.1.5.3), and aggregation of data (Sect. 8.2). All these solutions are summarized in this chapter.

8.1   Video Surveillance

8.1.1   Introduction

Instead of using traditional vision systems built essentially from fixed video cameras, it is possible to deploy autonomous and small wireless video sensor nodes (WVSN) [5] to achieve video surveillance of a given area of interest. Doing so lead to a much higher level of flexibility, therefore extending the range of surveillance applications that could be considered. More interestingly, this scenario can support dynamic deployment even in so-called object and obstacle-rich environments or hard-to-access areas. Such wireless video sensor nodes can in addition be thrown in mass to constitute a large scale surveillance infrastructure. In these scenarios, hundreds or thousands of video nodes of low capacity (resolution, processing, and storage) of a same or similar type can be deployed in an area of interest.

Surveillance applications have very specific needs due to their inherently critical nature associated with security [81, 108, 143]. The basic objective of video surveillance systems is to allow detection and/or identification of intruders. Therefore, in that context, the main goal of a video sensor network is to ensure the coverage of the whole area of interest at any time t. Another issue of prime importance is related to energy considerations since the scarcity of energy does have a direct impact on coverage, as it is not possible to have all the video nodes in activity at the same time. Therefore, a common approach is to define a subset of the deployed nodes to be active while the other nodes can sleep. There are already some techniques that schedule video nodes to work while maintaining the complete coverage [116,117,135]. The main idea in these techniques is to turn off a redundant node. Here redundancy means that the area surveilled by a node is completely covered by its neighbors too. However, these techniques usually depend on location or directional information, which is costly in energy and complexity. Usually it is very difficult to determine the redundant nodes without the location information. Fortunately, not all applications need a complete coverage at any time, and in most surveillance applications for intrusion detection, most sensor nodes can move to a so-called "idle mode" in the absence of intrusions. When an intruder is detected by a node all the network will be alerted. In that context, it is critical to provide an effective scheme for turning off video nodes without degrading the surveillance quality.

In this section, we present a solution to the joint scheduling problem in surveillance applications using video sensor nodes. We provide a chaotic sleeping scheme and conduct a theoretical and simulation analysis of both performances and security. Until research published in [23, 24], only random approaches had been extensively studied in the literature to turn off video nodes without degrading surveillance quality. Even if such methods present good scores in detecting random intrusions while preserving the lifetime of the network, *they do not encompass the situation of a malicious attacker.* That is to say, the intruder is not supposed to know something about the surveillance scheme, he cannot observe the WVSN for a while, or he is not authorized to deduce anything from his possible knowledge. In this presentation, we intend to tackle situations where the attacker is not supposed passive: he is smart and does not necessarily choose a random way to achieve his intrusion. We will show that chaos, as presented in the first parts of this book, is very useful to achieve security efficiently. In addition to preserving the network lifetime and being able to face random attacks, we show that our scheme is also capable of withstanding attacks of a malicious adversary due to its unpredictable behavior.

This section is organized as follows. In Section 8.1.2, related works in surveillance applications with WVSN are presented. Smart threats and malicious attackers are introduced in Section 8.1.3. The surveillance scheme, based on the chaos theory previously recalled, is detailed in Section 8.1.4. We show in Section 8.1.5 that the proposed scheme can be used against malicious attacks. Simulation results in Section 8.1.6 compare the scheme to a classical random schedule in terms of intruder's stealth time, network lifetime, and energy distribution. The section ends by a discussion, where our contribution is summed up and planned future work is detailed.

8.1.2 Related works

In video sensor networks, minimizing energy consumption and prolonging the system lifetime are major design objectives. Due to the significant energy savings when a node is sleeping, a frequently used mechanism is to schedule the sensor nodes such that redundant nodes go to sleep as often and for as long as possible. By selecting only a subset of nodes to be active and keeping the remaining nodes in a sleep state, the energy consumption of the network is reduced, thereby extending the operational lifetime of the WSN.

In this context, the coverage problem for wireless video sensor networks can be categorized as:

- *Known-Targets Coverage Problem,* which seeks to determine a subset of connected video nodes that covers a given set of target-locations scattered in a 2D plane.

- *Region-Coverage Problem,* which aims to find a subset of connected video nodes that ensures the coverage of the entire region of deployment in a 2D plane.

Most of the previous works have considered the known-targets coverage problem [4, 44, 53, 96]. The objective is to ensure at all times the coverage of some targets with known locations that are deployed in a two-dimensional plane. For example, the authors in [53] organize sensor nodes into mutually exclusive subsets that are activated successively, where the size of each subset is restricted, and not all of the targets need to be covered by the sensors in one subset. In [4], a directional sensor model is proposed, where a sensor is allowed to work in several directions. The idea behind this is to find a minimal set of directions that can cover the maximum number of targets. It is different from the approach described in [44] that aims to find a group of non-disjoint cover sets, each set covering all the targets to maximize the network lifetime.

Regarding the Region-Coverage Problem in which this study takes place, existing works focus on finding an efficient deployment pattern so that the average overlapping area of each sensor is bounded. The authors in [98] analyze new deployment strategies for satisfying some given coverage probability requirements with directional sensing models. A model of directed communications is introduced to ensure and repair the network connectivity. Based on a rotatable directional sensing approach, the authors in [130] present a method to deterministically estimate the amount of directional nodes for a given coverage rate. A sensing connected sub-graph accompanied with a convex hull method is introduced to model a directional sensor network into several parts in a distributed manner. With adjustable sensing directions, the coverage algorithm tries to minimize the overlapping sensing area of directional sensors only with local topology information. Last, in [117], the authors present a distributed algorithm that ensures both coverage of the deployment area and network connectivity, by providing multiple cover sets to manage Field of View redundancies and reduce objects' disambiguation.

All the above algorithms depend on the geographical location information (position and direction) of video nodes. These algorithms aim to provide a complete-coverage network so that any point in the target area would be covered by at least one video node. However, this strategy is not as energy-efficient as what we expect because of the following two reasons. First, the energy cost and system complexity involved in obtaining geometric information may compromise the effect of those algorithms. Second, video nodes located at the edge of the area of interest must be always in an active state as long as the region is required to be completely covered. These video nodes will die after some time, and their coverage area will be left without surveillance. Thus, the network coverage area will shrink gradually from outside to inside. This condition is unacceptable in video surveillance applications and intrusion detection, because the major goal here is to detect intruders as they cross a border or as they penetrate a protected area.

One way to solve these problems is to schedule a node to sleep following a probabilistic approach. Each node remains awake with a given probability so that the coverage of the area can be guaranteed. However, the probability can be modeled by an observer, who can take benefits from his observations to predict the dynamic of the network. This is obviously a security flaw. These considerations have led us to the introduction of smart threats in [24], which are recalled in the next section.

8.1.3 Smart threats

8.1.3.1 Presentation

Let us suppose that an adversary tries to reach a location X in the area without being detected. We consider that this situation leads to two categories of attacks against WVSN surveillance.

On the one hand, the attacker only knows that the area is under surveillance. He tries to take a chance, for example by following the shortest way or by trying a random path. In this first category of attacks that we call "blind elementary attacks," the intruder does not know how the surveillance is achieved as he does not observe the WVSN.

On the other hand, in the second category of attacks, called "malicious attacks" in this book, the intruder is supposed to be intelligent. He can try to take benefits from his observations to understand the behavior of the WVSN. After having recorded the dynamic of the WVSN for a given time, the malicious intruder can try to determine when video nodes are turned on. This prediction can help the intruder find a way to reach a location X without being detected.

In our opinion, the most reasonable way to evaluate the consequences of a malicious attack is to suppose that the intruder has access to the surveillance scheme. With this supposition, this security model encompasses the case where an attacker can have a physical access to a given node, thus determining the embedded mechanism used for video surveillance. In this Kerckhoffs-based principle, the attacker knows all but the initial parameters of the nodes. Moreover, he can observe the WVSN for a while. To achieve his intrusion, he can use all of the acquired knowledge; the sole difficulty is his lack of a secret parameter (the secret key) used to initialize the surveillance process.

The context of blind elementary attacks is well-known and understood: it has been studied a lot in the last decade, and various solutions have been proposed (Section 8.1.2). On the contrary, the case of an intelligent intruder (smart threat) has only been treated in [24]. In this chapter, we recall our proposal, to withstand attacks encompassing these malicious intrusions, and thus, offer a first solution to the problem raised by the smart threats existence hypothesis.

Technically speaking, the proposed approach offers several benefits. First, the node scheduling algorithm does not need location information. Therefore, the energy consumption is reduced because there is no need to locate the node itself and its neighbors. Second, we have shown in [24] that it performs as well as a random scheduling, in terms of lifetime and intrusion detection against blind elementary attacks (see Section 8.1.6). Finally, due to its chaotic properties, its coverage is unpredictable, and thus a malicious adversary has no solution to attack the network (Section 8.1.5).

8.1.3.2 Classification of malicious attacks

When a malicious adversary attacks a WVSN, he can concentrate his efforts either on the global network or on some specific nodes. Depending on the considered situation, he can perform either an active attack, modifying the network architecture or a node, or a passive attack based only on observations. He can have access to several WVSN using the same algorithm. Furthermore, he can build its own network to make some experiments. His objective is to find the secret key used in the targeted network; with this knowledge, the attacker will be able to predict the behavior of the video sensor nodes.

Active attacks have already been investigated several times in the literature. These studies encompass the cases where nodes can be added, moved, modified, or removed, where communications between nodes can be observed or modified, and where the global architecture of the network is attacked. However, some WVSN are such that any modification of the network is signaled, leading to the impossibility of such active attacks. On the contrary, passive observations and deductions of a malicious attacker are always possible. To the best of our knowledge, these threats have not yet been investigated, except in [24].

We have proposed the following classification for passive malicious attacks [24].

- In the **Target Only Attack (TOA)**, the adversary can only observe targeted networks.

- In the **Constant Key Attack (CKA)**, the adversary has access to several WVSN using the same secret key. The areas under surveillance and the network architecture change from one WVSN to another, but the attacker knows that all these networks use the same algorithm with the same secret key.

- In the **Known Original Attack (KOA)**, the attacker had previously accessed the WVSN and its area. He had the opportunity to test various keys in this previous access. He hopes that this knowledge will help him to determine a way to realize his intrusion when the WVSN is really launched.

- In the **Chosen Key Attack (CKA)**, the adversary has access to an exact copy of both the network and the area under surveillance that he wants to attack. He has realized for instance a miniature model or a computer simulator having exactly the same behavior as the targeted network and its area. He can thus try several secret keys, and if he manages to reproduce exactly the same behavior for the network, then he can reasonably suppose that the true secret key has been discovered.

- Finally, in the **Estimated Original Attack (EOA)**, the attacker has only an estimation, an approximation of the network and its area.

In each of these categories, the sole objective of the attacker is to obtain the value of the secret key. With this knowledge, he will be able to determine the WVSN behavior, finding a way to achieve his intrusion [23, 24].

8.1.3.3 Security levels in CKA

We now tackle the Chosen Key Attack problem. Let k_0 be the secret key used to initiate the video-surveillance. Denote by Y_k the probabilistic model that the attacker can build with his observations, and by $\mathbb{K}$ the set of all possible keys.

Definition 72 (Insecurity) The WVSN is insecure against the Target Only Attack if and only if $\exists k_1 \in \mathbb{K}, p(Y_{k_1}) = p(Y_{k_0})$ and $\forall k_2 \in \mathbb{K}, p(Y_{k_2}) \neq p(Y_{k_0})$

On the contrary,

Definition 73 (Security) The WVSN is secure against the Target Only Attack if and only if $\forall k \in \mathbb{K}, p(Y_k) = p(Y_{k_0})$

In that situation, it is easy to prove that the mutual information $\mathcal{I}(k_0, Y_{k_0})$ is equal to 0, which is often referred as *perfect secrecy*.

8.1.4 Chaos-based scheduling

8.1.4.1 Network capabilities

The WVSN is supposed to be constituted by 2^N nodes $V_i, i \in [\![0, 2^N - 1]\!]$. Each V_i is able to wake up on a specific signal, to survey a given area (and to detect intrusions), to send a wake up signal to another node V_j, and to go to sleep when it is required. Furthermore, it is supposed that V_i embeds:

- The mechanisms required by the intrusion detection: a sensing function $c_i(t)$, such as a camera, which returns some digital data at each listening time, and a decision function $d_i(c)$ that returns whether an intrusion is detected in the sensing values $(c_i(t))$ or not.

- An internal clock having the time $T_i = r_i T_0$ as a reference.

- A vector of N binary digits, called *the state of the system* V_i, and the capability to swap each bit of this vector $(0 \leftrightarrow 1)$.

- An integer e_i, called *listening time*, initialized to 0.

In other words, each node V_i can achieve CIs as they are defined in Chapter 5. Thus, each node can compute, easily and by using a few resources, a hash value and some pseudorandom numbers using tools introduced in the previous chapter. We will denote by g_i the seed of the PRNG used in node V_i, which is equal to a secret parameter p_i at time $t = 0$. This secret parameter with N bits has been generated by a cryptographically secure PRNG, and thus it is uniformly distributed into $[\![0; 2^N - 1]\!]$. The state V_i is initialized to the binary decomposition of g_i.

8.1.4.2   Deploying the network

The deployment of video sensor nodes in the physical environment is the first operation (step) in the network life cycle. It may take several forms.

- Sensor nodes may be randomly deployed by dropping them from a plane, and placed one by one by a human or a robot.

- Deployment may be a one-time activity or a continuous process.

These methods have been extensively studied in the literature [141]. In our method, the sole requirement to satisfy is to guarantee the uniform distribution into the region of interest.

8.1.4.3   Initialization of the WVSN

At time $t = 0$, a subset $\mathcal{I} \subset [\![0, 2^N - 1]\!]$ of nodes are woken up and $\forall i \in \mathcal{I}, e_i^{t_0} = T_i$.

8.1.4.4   Surveillance

The principle of the surveillance application proposed in [23, 24] can be summarized as follows. At each time $t_j = j \times T_0, j = 1, 2, \ldots$:

1. If a sleeping node V_i has received $n_i^{t_j-1} \geqslant 1$ wakeup orders during the time interval $[t_j-1, t_j]$, then it goes into active mode and sets its listening time $e_i^{t_j}$ to $n_i^{t_j-1} T_i$.

2. If an active node V_i has received $n_i^{t_j-1} \geqslant 1$ orders to wake up during the time interval $[t_j-1, t_j]$, then it increments its listening time: $e_i^{t_j} = e_i^{t_j-1} + n_i^{t_j-1} T_i$.

3. For each node V_i having a listening time $e_i^{t_j} \neq 0$:

- V_i ensures the surveillance of its area during T_0,
- If, during this time interval, an intrusion is detected, then the WVSN is under alert.
- If t_j is the first listening time of V_i after having activated, then:
  - The hash value $h_i^{t_j}$ of the sensed value $c_i(t_j)$ is computed (*c.f.* Section 5.1).
  - The seed g_i of the PRNG of V_i is set to $h_i^{t_j} + t_j$, where $+$ is the concatenation of the digits of $h_i^{t_j}$ and t_j (thus even if $h_i^{t_j} = h_i^{t_k}, k < j$, we have $g_i^{t_j} \neq g_i^{t_k}$).
  - The N bits of the state of the system V_i are set to $E_i^{t_j}$, where $E_i^{t_j}$ is the binary decomposition of i shown as a binary vector of length N.

4. N bits are computed with the PRNG of V_i. These bits define an integer $S_i^{t_j} \in [\![0, 2^N - 1]\!]$. Then the bit of $E_i^{t_j}$ in position $S_i^{t_j}$ is switched, which leads to a new state $E_i^{t_{j+1}}$. By doing so, CIs are realized.

5. Each active node V_i decreases its listening time: $e_i^{t_j} = e_i^{t_j} - 1$.

6. For each active node having its listening time $e_i^{t_j} = 0$:

  - V_i sends the wake up order to node V_k, where $k \in [\![0, 2^N - 1]\!]$ is the integer whose binary decomposition is the last state of the system V_i ($E_i^{t_{j+1}}$).
  - V_i goes to sleep.

8.1.5 Theoretical study

8.1.5.1 Scheduling as chaotic iterations

The scheduling scheme presented above can be described as CIs.

- The global state E^t of the whole system is constituted by the reunion of each internal state E_i^t of each node i. This is an element of $\mathbb{B}^{N \times 2^N}$.

- The strategy at time t is the subset of $[\![0; N \times 2^N]\!]$ constituted by all of the strategies that are computed into the awaken nodes at time t.

More precisely, if the node V_k has computed the strategy S_k^t at time t, then the global strategy S^t will contain the value $S_k^t + k \times N$. Last, the iteration function is the vectorial negation defined from $\mathbb{B}^{N \times 2^N}$ to $\mathbb{B}^{N \times 2^N}$. A subsequence E^{m^t} is extracted from E^t, which determines the changes that occur in the network: nodes whose binary id is into E^{m^t} are nodes that achieve the surveillance at the considered time. Let us remark that S^k and m^k depend both on the outside world, due to the fact that S_i^t are regularly seeded with the digest of some sensed values.

8.1.5.2 Complexity

Even if the hash function and the PRNG taken from [31] and [28] respectively can be replaced by any cryptographically secure hash function and PRNG, we do not recommend their substitution. Indeed, all of the operations used by our scheme can be achieved by CIs. Each iteration of CIs is only constituted by the negation of a few binary digits.

Obviously, such an operation is fast and does not consume a lot of energy. By doing so, we thus obtain an efficient video surveillance scheduling scheme compliant with WVSN requirements. Section 8.1.6 will detail more quantitatively this fact.

8.1.5.3 Coverage

The coverage of the whole area is guaranteed due to the following reasons.

First, the scheduling process corresponds to CIs. These iterations are chaotic according to Devaney, thus they are transitive. This transitivity property is the formulation of a uniform distribution in terms of topology. It claims that the system will never stop to visit any sub-region of the whole area, regardless of how tiny the region is.

Second, as the choice of the nodes to wake up at each time are done by using CIs, this selection corresponds to the returned value of our PRNG described in Section 7.2. Let us recall that this "$CI_{f_0}(X, Y) - generator$" takes two PRNGs X,Y as input sequences, realizes CIs with X as strategy, the vectorial negation as update function, and selects the states to publish as outputs by using the second PRNG Y (by such a combination, we improve the statistical properties of the input PRNG used as strategy, and we add chaotic properties). The scheduling process corresponds to the $CI_{f_0}(X, Y)$-generator, with X=m and Y=S. As Y is statistically perfect (Y is CI(ISAAC,ISAAC), which can pass the whole NIST, DieHard, and TestU01 batteries of tests), the uniform distribution of the states is then guaranteed.

Finally, experiments of Section 8.1.6 will show that this intended uniform coverage is well obtained in practice.

8.1.5.4 Security study

Qualitative Approach

Let us suppose that Oscar, an intruder, knows that the scheduling process is based on CIs, *i.e.*, he knows the whole algorithm, except the seeds that have been used to initiate the PRNGs of each node. By allowing that, we respect Kerckhoffs' principle: the adversary has all except the secret key. Oscar's desire is to reach a particular location X of the area without being detected. To achieve his goal, he can choose two strategies. On the one hand, he can try a blind elementary attack, either by following a random way from its position to X, of by choosing the shortest path. The next subsection and the experiments will show that such an attack cannot work. On the other hand, Oscar can try to take benefits both from his knowledge and his observations. However, if he can determine the nodes that are awake at time t, he cannot predict the awake nodes at time $t+1, t+2, \ldots$ To do so, he should be able to obtain $S^{t+1}, S^{t+2}, \ldots$, which are computed from the digests of some values that will be sensed in the future. As our hash function satisfies the avalanche effect (see Section 7.3), due to its chaotic properties, any error on the sensed value leads to a completely different digest.

As Oscar cannot determine the sensed values of each node, at each time, and with an infinite precision, he does not have the knowledge of the current state of the global system. He only has access to an approximation of this state. As the global scheduling process is chaotic, this error on the initial condition is magnified at each iteration, leading to the impossibility for Oscar to predict the scheduling process. This qualitative approach for security will be formalized in the remainder of this section.

Chaotic Properties

We now investigate the topological properties presented by the proposed video-surveillance scheme.

First of all, as proven in Chapter 7.1.5.3, chaotic iterations are expansive and topologically mixing when f is the vectorial negation f_0. Consequently, these properties are inherited by the WVSN presented previously, which induce a greater unpredictability. Any difference on the initial parameter of the WVSN is in particular magnified to be equal to the expansiveness constant.

Now, what are the consequences for a wireless sensor network to be chaotic according to Devaney's definition? First, the topological transitivity property implies indecomposability.

Hence, reducing the observed area in order to simplify its complexity, is impossible if $\Gamma(f)$ is strongly connected. Moreover, under this hypothesis the surveillance scheme is strongly transitive. According to this definition, for all pair of points x, y in the phase space, a point z can be found in the neighborhood of x such that one of its iterate $f^n(z)$ is y. Indeed, this result has been stated during the proof of the transitivity presented in 5. Among other things, the strong transitivity leads to the fact that without the knowledge of the initial awake nodes, all scheduling is possible. Additionally, no nodes of the output space can be discarded when studying the video-surveillance scheme; this space is intrinsically complicated and it cannot be decomposed or simplified.

Finally, these WVSN possess the instability property. This property, which is implied by sensitive point dependence on initial conditions, leads to the fact that in all neighborhoods of any point x, there are points that can be apart by ε in the future through iterations of the WVSN. Thus, we can claim that the behavior of these networks is unstable when $\Gamma(f)$ is strongly connected.

Cryptanalysis in CKA Framework

As stated in Section 8.1.5.1, the proposed video surveillance scheme can be rewritten as:

$$\begin{cases} X^0 \in \mathcal{X} \\ X^{k+1} = G_{f_0}(X^k), \end{cases} \tag{8.1}$$

where the phase space is $\mathcal{X} = [\![1; \mathsf{N} \times 2^\mathsf{N}]\!]^{\mathbb{N}} \times \mathbb{B}^{\mathsf{N} \times 2^\mathsf{N}}$, X^0 depends on a secret parameter $p = (p_1, \ldots, p_\mathsf{N}) \in (\mathbb{B}^\mathsf{N})^\mathsf{N}$ whose binary digits are uniformly distributed, and f_0 stands for the vectorial negation on $\mathbb{B}^{\mathsf{N} \times 2^\mathsf{N}}$.

We will now show that,

Proposition 38 *The video surveillance scheme proposed in this section is secure when facing a chosen key attack.*

PROOF Let $N = \mathsf{N} \times 2^\mathsf{N}$. We will prove by a mathematical induction that $\forall n \in \mathbb{N}, X^n \sim \mathbf{U}\left(\mathbb{B}^N\right)$.

The base case is immediate, as the initial state of the WVSN is initialized by $(g_1, \ldots, g_\mathsf{N})$, which are produced by a cryptographically secure PRNG, so $X^0 \sim \mathbf{U}\left(\mathbb{B}^N\right)$. Let us now suppose that the statement $X^n \sim \mathbf{U}\left(\mathbb{B}^N\right)$ holds for some n. Let $e \in \mathbb{B}^N$ and $\mathbf{B}_k = (0, \cdots, 0, 1, 0, \cdots, 0) \in \mathbb{B}^N$ (the digit 1 is in position k). So $P\left(X^{n+1} = e\right) = \sum_{k=1}^{N} P\left(X^n = e + \mathbf{B}_k, S^n = k\right)$. These two events are independent, thus:

$$P\left(X^{n+1} = e\right) = \sum_{k=1}^{N} P\left(X^n = e + \mathbf{B}_k\right) \times P\left(S^n = k\right).$$

According to the inductive hypothesis:

$$P\left(X^{n+1} = e\right) = \frac{1}{2^N} \sum_{k=1}^{N} P\left(S^n = k\right).$$

The set of events $\{S^n = k\}$ for $k \in [\![1; N]\!]$ is a partition of the universe of possible, so $\sum_{k=1}^{N} P\left(S^n = k\right) = 1$.

Finally, $P\left(X^{n+1} = e\right) = \frac{1}{2^N}$, which leads to $X^{n+1} \sim \mathbf{U}\left(\mathbb{B}^N\right)$. This result is true $\forall n \in \mathbb{N}$, we thus have proven that,

$$\forall p, Y_p = X^{N_0} \sim \mathbf{U}\left(\mathbb{B}^N\right).$$

So the video surveillance defined in this chapter is secure in CKA.

8.1.6   Simulation results

This section presents simulation results on comparing our chaotic approach to the standard C++ `rand()`-based approach with random intrusions. We use the OMNET++ simulation environment and the next node selection will either use chaotic iterations or the C++ `rand()` function (`rand() % 2^n`) to produce a random number between 0 and 2^n (for further information, see [24]).

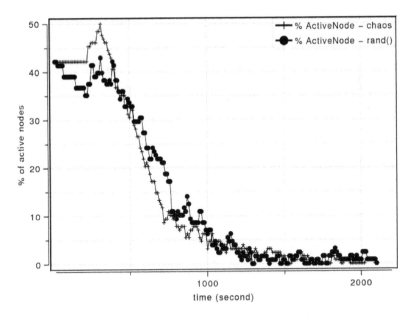

FIGURE 8.1: Percentage of active nodes.

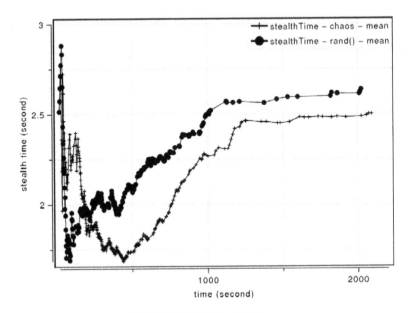

FIGURE 8.2: Stealth time.

For these sets of simulation, 128 sensor nodes (therefore $n = 7$) are randomly deployed in a $75m * 75m$ area. Unless specified, sensors have an 36° AoV and sensor node captures at the rate of 0.2fps. Each node starts with a battery level of 100 units and taking 1 picture consumes 1 unit of battery. When a node V_i is selected to wake up, it will be awake for T_i seconds. We set all $T_i = T = 20s$. According to the behavior defined in Section 8.1.4, before going to sleep after an activity period of $e_i T$, V_i will determine the next node to be waked up. It can potentially elect itself in which case V_i stays active for at least another T period. The elected node can be already active, in which case it simply increases its e_i counter. We set about 50% of the sensor nodes to be active initially (each sensor draws a random value between 0 and 1 and if the value is greater than 0.5, it will be active). This initial threshold is tunable but we did not try to vary this parameter in this chapter. The results presented here have been averaged over 10 simulation runs with different initial seeds. Figure 8.1 shows the percentage of active nodes. Both the chaotic and the standard `rand()` function have similar behavior: the percentage of active nodes progressively decreases due to battery shortage.

To compare both approaches in term of surveillance quality, we record to stealth time when intrusions are introduced in the area of interest. The stealth time is the time during which an intruder can travel in the field without being seen. The first intrusion starts at time 10s at a random position in the field. The scan line mobility model is then used with a constant velocity of 5m/s to make the intruder moving to the right part of the field. When the intruder is seen for the first time by a sensor, the stealth time is recorded and the mean stealth time computed. Then a new intrusion appears at another random position. This process is repeated until the simulation ends (*i.e.*, no more sensor nodes with energy). Figure 8.2 shows the mean stealth time over the whole simulation duration. Figure 8.3 presents the same data but with a sliding window averaging filter of 20 values. As the nodes are uniformly distributed in the area of interest, there is a strong correlation between the percentage of active nodes and the stealth time as it can be expected. The result we want to highlight here is that our chaotic node selection approach has a similar level of performance in presence of random intrusions as the standard `rand()` function while providing a formal proof of non-prediction by malicious intruders.

The last result we want to show is the energy consumption distribution. We recorded every 10s the energy level of each sensor node in the field and computed the mean and the standard deviation. Figure 8.4 shows the evolution of the standard deviation during the network lifetime. We can see that the chaotic node selection provides a slightly better distribution of activity than the standard `rand()` function.

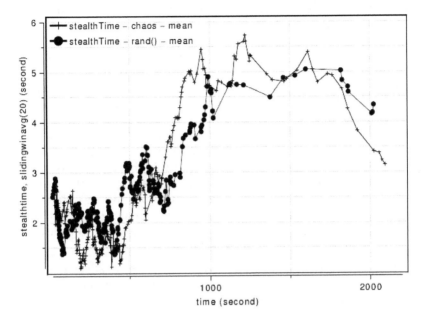

FIGURE 8.3: Stealth time.

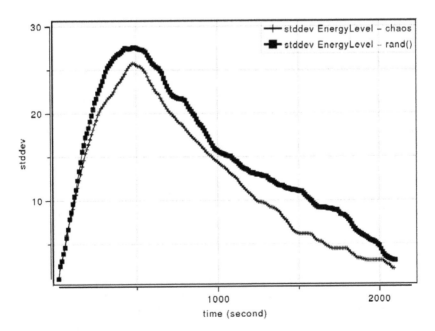

FIGURE 8.4: Evolution of the energy consumption's standard deviation.

8.1.7 Conclusion and perspectives of the chaos-based surveillance

In this section, a sleeping scheme for nodes has been proposed as an effective and secure solution to the joint scheduling problem in surveillance applications using WVSNs. It has been evaluated through theoretical and practical aspects of performance and security. As opposed to existing works, this scheduling scheme is not based only on randomness, but on the mathematical theory of chaos too, as presented in the first parts of this book. By doing so, we reinforce coverage and lifetime of the network, while obtaining a more secure scheme. We have considered in this chapter the case where the intruder is smart and active. Furthermore, we have supposed that he can know the scheme and observe the behavior of the network. We have shown that, in addition of being able to preserve WVSN lifetime and to present comparable results against random attacks, our scheme is also able to withstand such malicious attacks due to its unpredictable behavior.

In the next section, a second hot topic in WSN is investigated, namely the secure aggregation of data. It allows us to have a global approach to problems specific to security in wireless sensor networks.

8.2 Secure Aggregation

8.2.1 Presentation of the problem

As recalled before, a typical sensor network is expected to consist of a large number of sensor nodes deployed randomly on a large scale. Usually, these nodes have limited power, storage, communication, and processing capabilities, making energy consumption an issue.

A major functionality of a sensor node is to measure environmental values using embedded sensors, and transmit it to a base station called a "sink." The sensed data needs to be analyzed, which eventually serves to initiate some action. This analysis almost presumes computation of the maximum, minimum, average, *etc.* It can be either done at the base station or by the nodes themselves, in a hierarchical scenario. In order to reduce the amount of data to be transmitted to the sink, it is beneficial that this analysis can be done over the network itself. To save the overall energy resources of the network, it is agreed that the sensed data needs to be aggregated on the way to its final destination. Sensor nodes send their values to certain special nodes, i.e., aggregators. Each aggregator then condenses the data prior to sending it on. In terms of bandwidth and energy consumption, aggregation is beneficial as long as the aggregation process is not too central processing unit (CPU) intensive. The aggregators can either be special (more powerful) nodes or regular sensor nodes.

At the same time, sensor networks are often deployed in public or otherwise untrusted and even hostile environments, which prompts a number of security issues (e.g., key management, privacy, access control, authentication, *etc.*). Then, if security is necessary in other (e.g., wired or MANET) types of networks, it is much more so in sensor networks. Actually, it is one of the most popular research topics in this field and many advances have been reported in recent years.

From the above observations, we can notice the importance of a cooperative secure data aggregation in sensor networks. In other terms, after the data gathering and during transmissions to the base station, each node along the routing path cooperatively integrates and secures the fragments' messages. Therefore, secure data aggregation protocols require sensor nodes to encrypt or authenticate any sensed data prior to its transmission, implement data aggregation at every intermediate node (without decryption), and prefer data to be decrypted by the sink so that energy efficiency is maximized.

The benefit and vulnerability, as well as the need to secure in-network aggregation, have been identified by numerous schemes in the literature. One approach [118] proposed a secure information aggregation protocol to answer queries over the data acquired by the sensors. Even though their method provided data authentication to guarantee secrecy, the data was still sent in plain text format, which removes the privacy during transmission. Another one [45] proposed a secure energy efficient data aggregation (ESPDA) to prevent redundant data transmission in data aggregation. Unlike conventional techniques, their scheme prevents the redundant transmission from sensor nodes to the aggregator. Before transmitting sensed data, each sensor transmits a secure pattern to the aggregator. Only sensors with different data are allowed to transmit their data to the cluster-head. However, since each sensor at least needs to transmit a packet containing a pattern once, power cannot be significantly saved. In addition, each sensor node uses a fixed encryption key to encrypt data, which can lead to severe security flaws. In [83], the authors presented a secure encrypted-data aggregation scheme for wireless sensor networks. The idea is based on eliminating redundant sensor readings without using encryption and maintains data secrecy and privacy during transmission. This scheme saves energy on sensor nodes but still does not guarantee the privacy of sent data.

In [22], we have provided for the first time a hybrid approach for secure data aggregation in sensor networks, which completes the chaos-based approach for scheduling and coverage, leading to a truly secure WSN. First, our approach ensures that secrecy of sensed data is never disclosed to unauthorized parties, by providing a secure homomorphic cypher-system that allows efficient aggregation of encrypted data. We show that our encryption method allows many operations over cypher-texts that prevents data decryption at intermediate nodes (aggregators) and reduces energy consumption. Second, we extend our homomorphic secure data aggregation level to a two layer hierarchical data aggregation protocol, by including a watermarking-based authentication level. To assess the practicality of our technique, we evaluate it and compare it to existing cypher-systems. The obtained results show that we significantly reduce computation and communication overhead and our secure aggregation method can be practically implemented in on-the-shelf sensor platforms.

The remainder of this section is organized as follows. After having recalled some previous related work in the fields of data confidentiality and authentication, the next section introduces our two security layers that have been published in [21, 22]. In Section 8.2.4 the first one, namely the secure data aggregation using an almost fully homomorphic cryptosystem over elliptic curves, is presented in detail. Its security is evaluated qualitatively and through experiments in Section 8.2.4.6. In the next subsection, the last complementary approach for security in WSN is proposed. This is a node authentication protocol based on the work of Section 7.1 (in the information hiding security field). Advanced notions of security are taken from this field and translated in WSN terms. Then an existing authentication scheme is evaluated and improved in Section 8.2.5.2. Finally, in Section 8.2.5.3, a new secure authentication method based on watermarking is proposed and evaluated. This section ends with a discussion.

8.2.2   Security in sensor networks

Because sensor networks may interact with sensitive data and may be deployed in hostile unattended environments, it is imperative to protect sensitive information transmitted by sensor nodes. Moreover, wireless sensor networks introduce severe resource constraints due to their lack of data storage and power. Therefore, they have security problems that traditional networks (computer security) do not face and there are many security considerations that should be investigated.

In this section, we treat some essential security requirements that are raised in a wireless sensor network environment, mainly: data confidentiality, node authentication, and how they relate with the data aggregation process.

8.2.2.1   Data confidentiality

In critical applications, data confidentiality ensures that secrecy of transmitted data is never disclosed to unauthorized parties. Therefore, it is very important to build secure channels between sensor networks. The standard technique for keeping sensitive data secret is to encrypt them such that only intended receivers can realize decryption, hence achieving confidentiality.

Data encryption becomes necessary in sensor networks when this type of sensor can be subject of many types of attacks [52]. Without encryption, adversaries can monitor and inject false data into the network. In a general manner the encryption process is done as follows: sensor nodes must encrypt data on a hop-by-hop basis. An intermediate node (*i.e.*, aggregator) possessing the keys of all sending nodes, decrypts the received encrypted value, aggregates all received values, and encrypts the result for transmission to the base station. Though viable, this approach is fairly expensive and complicated, due to the fact of decrypting each received value before aggregation, which generates an overhead imposed by key management and prevents end-to-end data confidentiality.

Some privacy homomorphism based researches have been investigated recently [1,49,71]. They claim that, without participating in checking, the aggregators can directly aggregate the encrypted data. The problem of aggregating encrypted data in sensor networks was introduced in [71] and further refined in [49]. The authors propose to use homomorphic encryption schemes to enable arithmetic operations over cypher-texts that need to be transmitted in a multi-hop manner. However, these approaches provide a higher level of system security, since nodes would not be equipped with private keys, which would limit the advantage gained by an attacker compromising some of the nodes. Unfortunately, existing privacy homomorphisms used for data aggregation in sensor networks have exponential bounds in computation. For instance, Rivest Shamir Adleman (RSA) based cryptosystems [80, 95] are used, which require high CPU and memory capabilities to perform exponential operations. It is too computationally expensive to implement in sensor nodes. Moreover, the expansion in bit size during the transformation of plain text to cypher-text introduces costly communication overhead, which directly translates to a faster depletion of the sensors' energy. On the other hand, and from a security viewpoint, the cryptosystems [62] used in these approaches were cryptanalized [54,134], which means they cannot guarantee anymore high security levels.

In this section, we explain how to relax the statements above by investigating elliptic curve cryptography that allows feasible and suitable data aggregation in sensor networks beside the security of homomorphisms schemes. First of all, our scheme formerly proposed in [20] (and further investigated in [21, 22, 29]) for secure data aggregation in sensor networks, is based on a cryptosystem that has been proven safe and has not been cryptanalyzed. Indeed, it is known to be the sole secure and almost fully homomorphic cryptosystem usable now. Another property that enforces the security level of such an approach is coming from the fact that, as it is the case in ElGamal cryptosystem, for two identical messages it generates two different cryptograms. This property suggested fundamental for security in sensor networks [40,83,94] was not addressed in previous homomorphism-based security data aggregation researches. Beside all these properties and due to the use of elliptic curves, this proposal saves energy by allowing nodes to encrypt and aggregate data without the need of high computations. Last, the scheme we use allows more aggregation types over cypher data than the homomorphic cryptosystems used until now. This approach, which summarizes the authors' work in this domain [20–22,29], is detailed in Section 8.2.4 and is evaluated in Section 8.2.4.6.

8.2.2.2   Node authentication

In wireless sensor networks an adversary can change the whole packet stream by injecting additional packets. Therefore, the receiver needs to ensure that the data used in any decision-making process originates from the correct source. Without authentication, an adversary could masquerade as a node, thus gaining unauthorized access to resource and sensitive information, and interfering with the operation of other nodes. Moreover, a compromised node may send data to its data aggregator under several fake identities so that the integrity of the aggregated data is corrupted. Node authentication enables a sensor node to ensure the identity of the peer node it is communicating with. In the case of only two-nodes' communication, authentication can be achieved through a purely symmetric key cryptography: the sender and the receiver share a secret key to compute the message authentication code (MAC) of all communicated data.

In-network processing presents a critical challenge for data authentication in wireless sensor networks. Current schemes relying on MAC cannot provide natural support for this operation, because a MAC computation is a very energy-consuming operation. Additionally, even a slight modification to the data invalidates the MAC.

The authors in [114] propose a key-chain distribution system for their μTESLA secure broadcast protocol. The basic idea of the μTESLA system is to achieve asymmetric cryptography by delaying the disclosure of the symmetric keys. In this case a sender will broadcast a message generated with a secret key. After a certain period of time, the sender will disclose the secret key. The receiver is responsible for buffering the packet until the secret key has been disclosed. After disclosure the receiver can authenticate the packet, provided that the packet was received before the key was disclosed. One major limitation of μTESLA is that some initial information must be unicast to each sensor node before authentication of broadcast messages can begin.

In [142] a new way to achieve authentication through wireless sensor networks is introduced. It is based on digital watermarking and proposes an end-to-end, statistical approach for data authentication that provides inherent support for in-network processing. In this scheme, authentication information is modulated as a watermark and superimposed on the sensory data at the sensor nodes. The key idea formerly presented in [142] is to visualize the sensory data at a certain time snapshot as an image. Each sensor node is viewed as a pixel and its value corresponds to the gray level of the pixel. Due to this equivalence, information hiding techniques can be used to authenticate a wireless sensor network.

In some well-defined situations, the watermarked data can be aggregated by the intermediate nodes without incurring any *in route* checking. In this context, aggregation is for instance related to DCT or DWT compression, that is, to any operation over images that is able to reduce their weights without removing the watermarks. Upon reception of the sensory data, the sink is able to authenticate the data by finding and validating the watermark, thereby detecting whether the data has been illegitimately altered. In this way, the aggregation-survivable authentication information is only added at the sources and checked by the data sink, without any involvement of intermediate nodes. To realize such an authentication, the authors of [142] propose to use a data hiding scheme based on spread spectrum techniques. In their proposal, "each sensor node embeds part of the whole watermark into its sensory data, while leaving the heavy computational load of watermark detection at the sink." Moreover, as stated before, their scheme supports in-network aggregation. Such an approach, its issues and security consequences, and how to improve their scheme, is detailed in the remainder of this section.

8.2.3 Tree-based data aggregation

Data aggregation schemes aim to combine and summarize data packets of several sensor nodes so that the amount of data transmission is reduced. An example of data aggregation schemes is the tree based data aggregation protocol as presented in Figure 8.5 where sensor nodes collect information from a region of interest. When the user (sink) queries the network, instead of sending each sensor node's data to the base station, aggregators collect the information from its neighboring nodes, aggregates them, and sends the aggregated data to the base station over a multihop path.

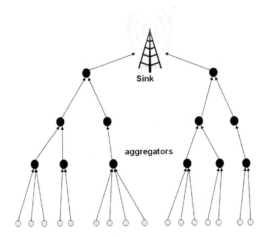

FIGURE 8.5: Tree-based data aggregation in sensor networks

The main objective of data aggregation is to increase the network lifetime by reducing the resource consumption of sensor nodes, especially the battery energy and bandwidth.

While increasing network lifetime, data aggregation protocols might take into account an important quality of service metric: the security. Therefore, encryption of the sensed data before its transmission becomes necessary and it is preferable to decrypt the data only at the base station level. In the next section, we present our model for sensor data encryption compliant with this requirement.

8.2.4   Sensor data encryption using fully homomorphic cryptosystem

In [20–22, 29] we were primarily concerned with data privacy in sensor networks. Our goal was to prevent attackers from gaining any information about sensor data. In that situation, ensuring end-to-end privacy between sensor nodes and the sink becomes problematic. This is largely because popular and existing cyphers are not additively homomorphic. In other words, the summation of encrypted data does not allow for the retrieval of the sum of the plain text values. Moreover, privacy existing homomorphisms usually have exponential bounds in computation.

To overcome this problem, in the model of [20–22, 29] we have proposed a security scheme for sensor networks using an elliptic curve based cryptosystem. We show here that this model permits many operations on encrypted data and does not demand high sensor capabilities and computation.

8.2.4.1   Operations over elliptic curves

In this section, we give a brief introduction to elliptic curve cryptography. The reader is referred to [79] for more details.

Elliptic curve cryptography (ECC) is an approach to public-key cryptography based on the algebraic structure of elliptic curves over finite fields [79]. Elliptic curves used in cryptography are typically defined over two types of finite fields: prime fields $\mathbb{F}_p$, where p is a large prime number, and binary extension fields $\mathbb{F}_{2^m}$ [79]. In this section, we focus on elliptic curves over $\mathbb{F}_p$.

Let $p > 3$, then an elliptic curve over $\mathbb{F}_p$ is defined by a cubic equation $y^2 = x^3 + ax + b$ as the set

$$\mathcal{E} = \left\{ (x, y) \in \mathbb{F}_p \times \mathbb{F}_p, y^2 \equiv x^3 + ax + b \ (\text{mod } p) \right\}$$

where $a, b \in \mathbb{F}_p$ are constants such that $4a^3 + 27b^2 \not\equiv 0 \ (\text{mod } p)$. An elliptic curve over $\mathbb{F}_p$ consists of the set of all pairs of affine coordinates (x, y) for $x, y \in \mathbb{F}_p$ that satisfy an equation of the above form and an infinity point $\mathcal{O}$.

The point addition and its special case, point doubling over $\mathcal{E}$ is defined as follows (the arithmetic operations are defined in $\mathbb{F}_p$) [79]. Let $P = (x_1, y_1)$ and $Q = (x_2, y_2)$ be two points of $\mathcal{E}$. Then:

$$P + Q = \begin{cases} \mathcal{O} & \text{if } x_2 = x_1 \text{ and } y_2 = -y_1, \\ (x_3, y_3) & \text{otherwise,} \end{cases} \qquad (8.2)$$

where:

Algorithm 7 Keys generation program in Python/Sage.

```
1: def GG1e(n):
2:      l = 1
3:      p = l*n-1
4:      while not isprime(p) or not p%3 == 2:
5:          l += 1
6:          p += n
7:      F = GF(p)
8:      H = EllipticCurve(F, [0, 1])
9:      X = H.gen(0)
10:     g = l*X
11:     G,y = [],g
12:     flag = True
13:     while flag:
14:         y = randint(0,n-1)*g
15:         if y.order() == n:
16:             G.append(y)
17:             if len(G) == 2:
18:                 flag = False
19:     return G,p
20:
21: def G(t):
22:     q1 = generatePrime(t)
23:     q2 = generatePrime(t)
24:     n = q1*q2
25:     GG,p = GG1e(n)
26:     return (q1,q2,GG,p)
27:
28: def KeyGen(bits):
29:     (q1,q2,GG,p) = G(bits)
30:     n = q1*q2
31:     g,u = GG
32:     h = q2*u
33:     return ((n,G,g,h,p),q1)
```

- $x_3 = \lambda^2 - x_1 - x_2,$

- $y_3 = \lambda \times (x_1 - x_3) - y_1,$

$$\lambda = \begin{cases} (y_2 - y_1) \times (x_2 - x_1)^{-1} & \text{if } P \neq Q, \\ (3x_1^2 + a) \times (2y_1)^{-1} & \text{if } P = Q. \end{cases} \tag{8.3}$$

Finally, we define $P + \mathcal{O} = \mathcal{O} + P = P, \forall P \in \mathcal{E}$, which leads to an abelian group $(\mathcal{E}, +)$. On the other hand the multiplication $n \times P$ means $P+P+....+P$ n times and $-P$ is the symmetric of P for the group law $+$ defined above for all $P \in \mathcal{E}$.

8.2.4.2   Public/private keys generation with ECC

In this section we show how we can generate the public and private keys for encryption, following the cryptosystem proposed by Boneh *et al.* [40]. The analysis of the complexity will be treated in a later section.

Let $\tau > 0$ be an integer called "security parameter." To generate public and private keys, first of all, two τ-bits prime numbers must be computed. Therefore, a cryptographic pseudorandom generator can be used to obtain two vectors of τ bits, q_1 and q_2 (see Section 7.2 for further details). Then, a Miller-Rabin test can be applied for testing the primality or not of q_1 and q_2. We denote by n the product of q_1 and q_2, $n = q_1 q_2$, and by l the smallest positive integer such that $p = l \times n - 1$. l is a prime number while $p = 2 \pmod 3$.

In order to find the private and public keys, we define a group H, which presents the points of the super-singular elliptic curve $y^2 = x^3 + 1$ defined over $\mathbb{F}_p$. It consists of $p + 1 = n \times l$ points, and thus has a subgroup of order n, we call it G. In another step, we compute g and u as two generators of G and $h = q_2 \times u$. Then, following [40], the public key will be presented by (n, G, g, h) and the private key by q_1.

To illustrate such a key's generation, we have proposed in [22] a program given in Algorithm 7. It is written with the Python 2.6 language and the Sage library to manipulate elliptic curves. The *randint(a,b)* function is provided by the random library; it generates an integer randomly picked into the interval $[a, b]$. The *generatePrime(n)* function is not detailed here. It receives an integer n as its input argument and generates a prime number of n bits.

8.2.4.3   Encryption and decryption

After the private/public keys' generation, we proceed now to the two encryption and decryption phases [79]:

- **Encryption :** Assuming that the messages space consists of integers in the set $\{0, 1, ..., T\}$, where $T < q_2$, and m the (integer) message to encrypt. First, a random positive integer is picked from the interval $[0, n - 1]$. Then, the cypher-text is defined by

$$C = m \times g + r \times h \in G,$$

  in which $+$ and $\times$ refer to the addition and multiplication laws defined previously.

- **Decryption :** Once the message C arrives at the destination, to decrypt it, we use the private key q_1 and the discrete logarithm of $(q_1 \times C)$ base $q_1 \times g$ as follows:

$$m = \log_{q_1 \times g} q_1 \times C.$$

This takes expected time $\sqrt{T}$ using the well-known Pollard's lambda method [79]. Moreover, this decryption can be sped-up by precomputing a table of powers of $q_1 \times g$.

Algorithm 8 Python program for Encryption and Decryption

```
1: def Encrypt(Kp, M):
2:     (n,G,g,h,p) = Kp
3:     r = randint(0,n-1)
4:     return M*g+r*h
5:
6: def Decrypt(Kp,Ks,C):
7:     (n,G,g,h,p), q1 = Kp, Ks
8:     P = q1*g
9:     return P.discrete-log(q1*C)
10:
11: def Decrypt-product(Kp,Ks,C):
12:     (n,G,g,h,p), q1 = Kp, Ks
13:     g1 = modified-weil(g,g,p)
14:     return log(C,g1)
```

In Algorithm 8 is detailed an example of encryption and decryption programs in Python/Sage. The *modified_weil* and *discrete_log* functions are provided by Sage.

8.2.4.4 Homomorphic properties

As mentioned before, the approach presented in [20–22, 29] ensures easy encryption/decryption without any need of extra resources. This assessment will be detailed in the next section. Moreover, this approach supports homomorphic properties, which gives us the ability to execute operations on values even though they have been encrypted. Indeed, it allows N additions and one multiplication directly on cryptograms, which prevents the decryption phase at the aggregators level and saves nodes energy, which is crucial in sensor networks.

Additions over cypher-texts are done as follows: let m_1 and m_2 be two messages and C_1, C_2 their cypher-texts respectively. Then the sum of C_1 and C_2, let us call it C, is represented by

$$C = C_1 + C_2 + r \times h,$$

where r is an integer randomly chosen in $[0, n-1]$ and $h = q_2 \times u$ as presented in the previous section. This sum operation guarantees that the decryption value of C is the sum $m_1 + m_2$. The addition operation can be done several times, which means we can do sums of encrypted sums.

The multiplication of two encrypted values and its decryption are done as follows: let e be the modified Weil pairing on the curve and g, h the points of G as defined previously. Let us recall that this modified Weil pairing e is obtained from the Weil pairing E [39,40] by the formula: $e(P,Q) = E(x \times P, Q)$, where x is a root of $X^3 - 1$ on $\mathbb{F}_{p^2}$. Then, the result of the multiplication of two encrypted messages C_1, C_2 is given by [40]

$$C_m = e(C_1, C_2) + r \times h_1,$$

where $h_1 = e(g,h)$ and r is a random integer pick in $[1,n]$.

The decryption of C_m is equal to the discrete logarithm of $q_1 \times C_m$ to the base $q_1 \times g_1$:

$$m_1 m_2 = \log_{q_1 * g_1}(q_1 \times C_m.)$$

where $g_1 = e(g,g)$.

The decryption program of a product is given in Algorithm 8 whereas the addition and multiplication over cryptograms' programs are given in Algorithm 9.

Algorithm 9 Python/Sage program of homomorphic operations

```
1: def multiply(Kp,cr1,cr2):
2:     (n,G,g,h,p) = Kp
3:     r = randint(0,n-1)
4:     return modified-weil(cr1,cr2,p)
5:          +r*modified-weil(g,h,p)
6:
7: def add(Kp,cr1,cr2):
8:     (n,G,g,h,p) = Kp
9:     r = randint(0,n-1)
10:    return cr1+cr2+r*h
```

8.2.4.5   Encryption for sensor networks

In previous secure data aggregation protocols, security and data aggregation are almost always achieved together in a hop-by-hop manner. That is, data aggregators must decrypt every message they receive, aggregate the messages according to the corresponding aggregation function, and encrypt the aggregation result before forwarding it. Therefore, these techniques cannot provide data confidentiality at data aggregators, and result in latency because of the decryption/encryption process.

In [20–22,29], we proposed an encryption protocol that performs data aggregation without requiring the decryption of the sensor data at data aggregators. We adopt the following scenario as shown in Figure 8.6: after collecting information, each sensor node encrypts its data according to elliptic curve encryption (*c.f.* Section 8.2.4.1) and sends it to the nearest aggregator. Then, aggregators aggregate the received encrypted data (without decryption) and send it to the base station, which in turn decrypts the data and aggregates it. We notice that all aggregators can do N additions and the final layer of aggregators can do one multiplication on encrypted data. For instance, aggregators can compute securely the following values.

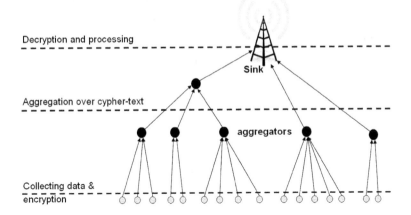

FIGURE 8.6: Secure data aggregation in sensor networks

- **Arithmetic Mean**

  To compute the average of nodes' measurements, aggregators can calculate the sum of the encrypted measurements and the number of nodes, and send it to the base station. More precisely, when using our scheme, each sensor encrypts its data x_i to obtain cx_i. The sensor then forwards cx_i to its parent, who aggregates all the cx_j's of its k children by simply adding them up. The resulting value and the encryption of k are then forwarded. The sink can thus compute the average value with all of these data.

- **Variance**

  Another common aggregation is to estimate the variance of the sensed values. Let us recall that the variance of n values $x_1, ..., x_n$ is defined by:

$$s_n^2 = \frac{1}{n} \sum_{i=1}^{n} (x_i - \bar{x})^2 = \left(\frac{1}{n} \sum_{i=1}^{n} x_i^2 \right) - \bar{x}^2.$$

  Our scheme can also be used to derive the variance of the measured and encrypted data, by the same method as in [48]. In this case, each sensor i must compute $y_i = x_i^2$, where x_i is the measured sample, and encrypts y_i to obtain cy_i. x_i must also be encrypted, as explained in the previous section. The sensor forwards cy_i, together with cx_i, to its parent. The parent aggregates all the cy_j of its k children by simply adding them up. It also aggregates, separately, the cx_j, as explained in the previous section. The two resulting values are then forwarded. The sink ends up with values $Cx = \sum_{i=1}^{n} cx_i$ and $Cy = \sum_{i=1}^{n} cy_i$. Cx is used to compute the average Av, when Cy is used to compute the variance as follows: $Var = \frac{Vy}{n} - Av^2$, where Vy is the decryption of Cy.

- **Weighted Mean**

    The weighted mean of a non-empty set of data $x_1, x_2, \ldots, x_n$ with non-negative weights $w_1, w_2, \ldots, w_n$, is the quantity

    $$\bar{x} = \frac{w_1 x_1 + w_2 x_2 + \cdots + w_n x_n}{w_1 + w_2 + \cdots + w_n}.$$

    We suppose now that each aggregator i of the first aggregation layer has computed the mean x_i of the encrypted values received from its sensor node. Additionally, we suppose that these aggregators are weighted, depending on their importance. For security reasons, this weight is also encrypted and the cypher value is denoted by w_i. This w_i can be proportional to the number of aggregated sensors. This weight can also illustrate the fact that two given regions have not the same importance. To achieve weighted mean, each aggregator multiplies its encrypted mean x_i with encrypted weight w_i as it has been explained previously. The resulting value is then forwarded to the sink, which can decrypt $w_i \times x_i$ and sum all these decrypted values, to obtain the weighted mean defined above.

8.2.4.6   Practical issues

In this section we present some practical issues to the data encryption model proposed in our previous research works. First, we study the sizes of the encryption keys and we compare it to existing approaches. Then, we show how we can optimize the sizes of cryptograms in order to save more sensors' energy.

Sizes of the Keys

Cryptograms are points of the elliptic curve $\mathcal{E}$. They are constituted by couples of integer coordinates lesser than or equal to $p = lq_1 q_2 - 1$.

It is commonly accepted [36, 91] that for being secure until 2020, a cryptosystem:

- must have $p \approx 2^{161}$, for EC systems over $\mathbb{F}_p$,

- must satisfy $p \approx 2^{1881}$ for classical asymmetric systems, such as RSA or ElGamal on $\mathbb{F}_p$.

Thus, for the same level of security, using elliptic curve cryptography does not demand high key sizes, contrary to the case of RSA or ElGamal on $\mathbb{F}_p$. The use of small keys leads to small cryptograms and fast operations for encryption.

Reducing the Size of Cryptograms

In this section is shown how we can reduce the size of cryptograms while using ECC. This is beneficial for sensor nodes in terms of reducing energy consumption by sending data with smaller sizes. The messages are encrypted with q_2 bits, which leads to cryptograms with a mean of 160 bits long.

Let us suppose that $p \equiv 3 \ (\bmod \ 4)$. As the cryptogram is an element (x, y) of $\mathcal{E}$, which is defined by $y^2 = x^3 + 1$, we can compress this cryptogram (x, y) to $(x, y \bmod 2))$ before sending it to the aggregator (as the value of y^2 is known). In this situation, we obtain cryptograms with a mean of 81 bits long for messages between 20 and 40 bits long.

To decompress the cryptogram (x, i), the aggregator must compute $z = x^3 + 1 \bmod p$ and $y = \sqrt{z} \bmod p$, which can be written as $y = z^{(p+1)/4} \bmod p$, then:

- if $y \equiv i (\bmod \ 2)$, then the decompression of (x, i) is (x, y),

- else the decompression point is $(x, p - y)$.

8.2.4.7 Security study

Due to hostile environments and unique characteristics of sensor networks, it is a challenging task to protect sensitive information transmitted by nodes to the end user. In addition, this type of network has security problems that traditional networks do not face. In this section, we outline a security study dedicated to wireless sensor networks' secured communications.

In a sensor network environment adversaries can commonly use the following attacks:

Known-plain text attack: They can use common key encryption to see when two readings are identical. By using nearby sensors under control, attackers can conduct a known-plain text attack.

Chosen-plain text attack: Attackers can tamper with sensors to force them to predetermined values.

Man-in-the-middle: They can inject false readings or resend logged readings from legitimate sensor motes to manipulate the data aggregation process.

In Tables 8.1, 8.2, and similar to [55], we present a comparison between different encryption policies and possible attacks. In our method, as data are encrypted by public keys, and these public keys are sent by the sink to the sole authenticated motes, the wireless sensor network is then not vulnerable to a Man-in-the-middle attack. Moreover, our approach guarantees that for two similar texts two different cryptograms are given, which prevents the Chosen-plain text attacks and the Man-in-the-middle attacks. Finally, as the proposed scheme possesses the homomorphic property, data aggregation is done without decryption, and no private key is used in the network.

TABLE 8.1: Encryption policies and vulnerabilities

Encryption Policy	Possible attacks
Sensors transmit readings without encryption	Man-in-the-middle
Sensors transmit encrypted readings with permanent keys	Known-plain text attack Chosen-plain text attack Man-in-the-middle
Sensors transmit encrypted readings with dynamic keys	None of above
Our scheme	None of above

TABLE 8.2: Encryption policies and aggregation

Encryption Policy	Data aggregation
Sensors transmit readings without encryption	Generating wrong aggregated results
Sensors transmit encrypted readings with permanent keys	Data aggregation is impossible, unless the aggregator has encryption keys
Sensors transmit encrypted readings with dynamic keys	Data aggregation cannot be achieved unless the aggregator has encryption keys
Our scheme	Data aggregation can be achieved

8.2.4.8   Experimental results

To show the effectiveness of this approach we conducted in [20–22, 29] a series of simulations comparing this method to another existing one based on RSA cryptosystem. We considered a network formed of 500 sensor nodes, each one equipped by a battery of 100 units capacity. We consider that the energy consumption "E" of a node is proportional to the computational time t, *i.e.*, $E = kt$. The same coefficient of proportionality k is taken while comparing the two encryption scenarii. Sensor nodes are then connected to 50 aggregators chosen randomly. Each sensor node chooses the nearest aggregator. The running of each simulation is as follows: each sensor node takes a random value, encrypts it using one of the encryption methods, then sends it to its aggregator. Aggregators compute the sum of the encrypted received data and send it to the sink. We compared our approach to the known RSA public-key cryptographic algorithms, and we evaluated the energy consumption of the network while varying the sizes of the keys and obviously the security lev-

TABLE 8.3: Security versus energy at the nodes level using our approach

Security level	Size p of the key	E (battery units)
1	46	0.02
2	85	0.05%
3	125	0.07
4	167	0.10

TABLE 8.4: Security versus energy at the nodes level using RSA

Security level	Size of the key	E (battery units)
1	472	0.08
2	945	0.53
3	1416	1.63
4	1891	3.63

els [20–22, 29]. The energy consumption is the units of the battery used to do the encryption.

Tables 8.3 and 8.4 show the energy consumption of sensor nodes used to do the encryption operations using our encryption method and the RSA one respectively. We varied the keys sizes and obviously the security levels. A security level is just an indicative factor of security (level 4 provides higher security than level 1). We notice that, for the same level of security, in the proposed approach we used smaller keys while saving more energy. For instance, for high security levels (4 for example) a node using our approach needs to use a key of 167 bits instead of 1891 in the case of RSA, and it consumes 0.1 % of the battery power instead of 3.63 %.

Tables 8.5 and 8.6 give the energy consumption E at the aggregation stage. The same hypotheses as above have been done; the sole difference is that aggregator nodes have a battery of 1000 units of energy. It can be seen that the energy needed by aggregators is between 50 and 500 times more important in the RSA-based scheme, for the same level of security.

Figure 8.7 gives the comparison between RSA and elliptic curve based encryption, concerning the average energy consumption of an aggregating wireless sensor network. We notice that our approach saves energy largely greater than the case of RSA, where its depletion is so fast. Finally let us notice that,

TABLE 8.5: Security versus energy at the aggregator level using our approach

Security level	Size p of the key	E (battery units)
1	46	0.02
2	85	0.04
3	125	0.07
4	167	0.10

TABLE 8.6: Security versus energy at the aggregator level using RSA

Security level	Size of the key	E (battery units)
1	472	1.13
2	945	8.09
3	1416	24.74
4	1891	56.27

in addition to reducing the amount of energy units needed for encryption and aggregation, the sink receives many more values per second in EC-based networks than in the RSA-based one.

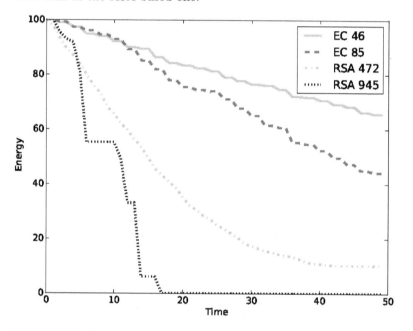

FIGURE 8.7: Comparison of energy consumption in aggregator nodes

8.2.5   Authentication over homomorphic sensor networks

In the previous sections, we have proposed to use a homomorphism encryption scheme to support in-network processing while preserving privacy. Compared to existing secure aggregation schemes based on homomorphism encryption, our method has not been cryptanalysed. Moreover, due to the possibility to realize n additions and one product over the cypher values, this scheme enlarges the variety of allowing aggregation operations through cyphertexts.

However, all of the secure homomorphism encryption schemes only allow some specific query-based aggregation functions, *e.g.*, sum, average, *etc.* Indeed data encryption guarantees that only intended parties obtain the unencrypted plain data, it does not protect the network from malicious or spoofed packets. Node authentication enables a sensor node to ensure the identity of the packet's sender. Another way to achieve secure data aggregation in wireless sensor networks is then to authenticate sensing values.

Finally, an hybrid approach of secure data aggregation in wireless sensor networks can be obtained by combining homomorphic encryption and watermarking-based authentication. In the next section we present a scheme formerly proposed in [21, 22], aiming at authenticating nodes in sensor networks.

8.2.5.1 Information hiding-based authentication

In this section, a connection between authentication in wireless sensor networks and digital watermarking, as presented in Section 7.1, is provided: we consider that authentication information is modulated as a watermark and superposed on the sensory data at the sensor nodes. The watermarked data can be aggregated by the intermediate nodes without incurring any en route checking. Upon reception of the sensory data, the data sink is able to authenticate the data by validating the watermark, thereby detecting whether the data has been illegitimately altered.

Let us explain why, in our opinion, the information hiding framework recalled in a previous chapter is useful for studying wireless sensor network authentication nodes.

Robustness for authenticated wireless sensor networks It is possible to reformulate the problematic of nodes authentication using information hiding techniques. First of all, robustness means that the watermark still remains after geometric and frequency attacks. The interest to have a robust watermarking for authentication in WSN is then twofold.

On the one hand, the network is not always fixed and can possibly evolve over time. Nodes can be moved for various reasons, some of them can stop to transmit their sensed and watermarked data (for technical reasons, or when they have consumed all of their energies), noise can appear during transmission, and so on. Nevertheless, the authentication capability of the whole network must be preserved in the sink, and thus the watermarking scheme used for authentication must be compliant with such alterations. That is to say, if the wireless sensor network can be slightly altered for reasonable and natural reasons, then the authentication scheme must be robust. Table 8.7 gives some relationships between geometric and frequency attacks in the information hiding framework, and natural alteration of a wireless sensor network.

On the other hand, some information hiding schemes are robust against image compression attacks, like JPEG or JPEG2000 compressions. Such

TABLE 8.7: Relationship between digital watermarking and WSN

Digital Watermarking	WSN
pixel	node
picture	network
zeroing attack	death of nodes
rotation and resize attacks	nodes displacement
uniform or gaussian noise	transmission errors
contrast attacks	unbalanced signals
blur attacks	signal attenuation

a resistance is obtained for instance by inserting the watermark into the DCT or DWT coefficients of the image instead of using the gray level of each pixel. The idea formerly proposed by [142] and extended in [21, 22] is to use this resistance for aggregation. Indeed, in situation of JPEG or JPEG2000 compression resistance, watermarked data sent by sensor nodes can be aggregated by using a DCT or DWT compression. Due to the robustness of the well-chosen information hiding scheme against these attacks, the watermark still remains after such compressions, and the aggregation preserves authentication.

Furthermore, a fragile watermarking can be useful too in an information hiding based authentication of nodes in a WSN. Let us consider for instance that an attacker adds one of his node in a given wireless sensor network that uses a fragile watermarking scheme for authentication. Additionally, we suppose that he can send corrupted sensed values without being detected, either to the aggregation layer or to the sink. Under this situation, the "watermarked image" received by the sink will be such that at least one pixel (*i.e.*, the corrupted node) has an incorrect piece of watermark. Due to the fragility of the scheme, this alteration will be magnified and the extracted watermark will be completely different from what was expected, leading to the detection of the attack. Such a fragile authentication scheme can be useful too in situations where an attacker tries to modify an authorized node. In this case, as the node embeds different pieces of watermark, the sink will be able to detect such an anomaly.

Finally, there exist some watermarking schemes that are fragile in almost all situations, but are robust against some well defined threats. So fragile and robust properties can help to determine the best watermarking scheme for a given WSN authentication context. For instance, some schemes can be found in the literature that are robust against JPEG attacks with small compression rate, but are fragile in all of the other types of attacks. Such an algorithm is helpful when the network cannot evolve geographically, must use compression based aggregation, and is in a hostile environment. As a conclusion, we can see that using a watermarking scheme for authentication through WSN is useful in many situations.

Security for authenticated WSN For robustness, information hiding security can be useful when authenticating nodes in wireless sensor networks. The four classes of attacks presented in Section 7.1 can be translated to WSN security as follows.

Watermark-Only Attack (WOA) occurs when an attacker has only access to several transmitted authenticated data. That is to say, he can only observe transmissions.

Known-Message Attack (KMA) occurs when an attacker has access to several pairs of watermarked contents and corresponding hidden messages. In other words, the adversary can observe the transmissions, and has found a way to insert an altered node into the attacker's watermark.

Known-Original Attack (KOA) is when an attacker has access to several pairs of watermarked contents and their corresponding original versions. That is to say, the attacker can determine which value a node has sensed and can see the resulting watermarked data sent by this node.

Constant-Message Attack (CMA) occurs when the attacker observes several watermarked contents and only knows that the unknown hidden message is the same in all contents. In that situation, the adversary can observe the transmitted data another time. Additionally, he knows that the same watermark is always used to authenticate these data.

Other categories of attacks can be found in the literature, such as the Estimated-Original Attack. They all can be translated into the wireless sensor network security framework.

The Simmons' prisoner problem put into the WSN context can be translated too, as follows. Eve observes the transmission between two nodes called Alice and Bob. She tries to determine whether a given transmission is authenticated or not. Obviously, this is for her the starting point of an attack to authentication. For instance, if she is able to spot the difference between authenticated and unauthenticated data, then she can:

- Replace unauthenticated data with her own values without being detected.
- Concentrate her efforts on a subset of authenticated data.

- Try to understand the differences between authenticated and unauthenticated data, with a view to forging her own "authenticated" values.

- Try to determine, by using statistical models and tools, the embedding key (the piece of watermark used to authenticate these values).

- *etc.*

The stego-security of Definition 65 means that such a separation between authenticated and unauthenticated data is impossible, as the use of any key does not change the probabilistic model of the transmitted data. Obviously, there is a lack of security if an authentication scheme of data sending through a WSN, with an adversary being able to observe transmissions (Simmons' prisoner problem, WOA setup), is based on a watermarking algorithm that is not stego-secure. Similar conclusions can be obtained with the chaos-security notion in the KOA, KMA, and CMA setups; if situations covered by these setups can possibly occur, then the watermarking scheme used for authentication must be chaos-secure.

To the best of our knowledge, until now, only two data hiding schemes have been used to authenticate data sending through a WSN. The first one is a spread-spectrum technique, used in [142]. The second one uses chaotic iterations [21, 22]. In what follows, these information hiding techniques are recalled and their security is evaluated.

8.2.5.2   Security study of Zhang *et al.* authentication scheme

Zhang *et al.* nodes authentication scheme for WSN is simply an authentication that uses the spread-spectrum data hiding, as we have recalled it in Section 7.1.4.1. Let us recall that natural watermarking has been proven stego-secure when $\eta = 1$, whereas all of the other spread-spectrum techniques are not stego-secure (see [50]). Additionally, this scheme is reputed to be not robust. Finally, these four techniques are chaos-secure so they can be considered when facing an attacker in the CMA context [78]. However, as these techniques are not expansive, they are unable to face an attacker in KOA and KMA setups [78]. We now propose a cryptanalysis of the Zhang *et al.* authentication scheme.

As reminded in Section 7.1, spread spectrum is known to be not robust: even if the Zhang *et al.* scheme survives to a certain degree of distortion, it cannot face an elementary blind attack. Furthermore, spread spectrum data hiding techniques are only stego-secure in the "Natural Watermarking" situation [50]. The spread-spectrum subclass used in [142] is related to classical SS, *i.e.*, with BPSK modulation [50]. This subclass is neither stego-secure [50], nor chaos-secure [30]. These lacks of security allow an attacker who observes the network to access to the secret embedding key in all of the following situations: WOA, KMA, KOA, and CMA setups.

To improve the security of the network in WOA setup, the use of Natural Watermarking instead of BPSK modulation is required [50]. However, we have shown in Section 7.1.4.1 that Natural Watermarking is less chaos-secure than the data hiding algorithm presented in [16]. This algorithm, based on chaotic iterations, is able to withstand attacks in KMA, KOA, and CMA setups [73]. Moreover, this technique is more robust than spread-spectrum, as it is stated in [18]. To sum up, the use of the scheme proposed in [16] improves the security and robustness of the scheme presented in [142]. This algorithm is recalled in the next section and evaluated in the last one.

8.2.5.3 Authentication based on chaotic iterations

For easy understanding, the explanation of the authentication scheme based on chaotic iteration is illustrated by using pictures instead of networks. By doing so, we obtain a scheme in WSN having a formulation similar to the information hiding algorithm presented in Section 7.1.5. As there is an equivalency between pixels and nodes, the discussion and the evaluation of Sect. 7.1.5 holds for a wireless sensor network, *mutatis mutandis*.

To explain how to use chaotic iterations for node authentication, we must first define the significance of a given node using the WSN ↔ picture equivalency. By doing so, we will have an authentication scheme whose formulation is similar to the information hiding algorithm presented in Section 7.1.5.1.

We first notice that in each node, the alteration of the sensed value for authentication must not be important. That is to say, terms of the original content x that may be replaced by terms issued from the watermark y are less important than others; they could be changed without being perceived as such. More generally and similarly to Section 7.1.5.1, a *signification function* attaches a weight to each sensed value, depending on its position t.

Definition 74 A *signification function* is a real sequence $(u^k)^{k \in \mathbb{N}}$.

Using this function, it is possible to define the significance of a given sensed value in a WSN.

Definition 75 Let $(u^k)^{k \in \mathbb{N}}$ be a signification function, m and M be two reals s.t. $m < M$.

- The *most significant coefficients (MSCs)* of the sensed values x of a WSN is the finite vector

$$u_M = \left(k \mid k \in \mathbb{N} \text{ and } u^k \geqslant M \text{ and } k \leq |x| \right);$$

- The *least significant coefficients (LSCs)* of x is the finite vector

$$u_m = \left(k \mid k \in \mathbb{N} \text{ and } u^k \leq m \text{ and } k \leq |x| \right);$$

- The *passive coefficients* of x is the finite vector

$$u_p = \left(k \mid k \in \mathbb{N} \text{ and } u^k \in]m; M[\text{ and } k \leq |x| \right).$$

For a given WSN x, MSCs are then ranks of x that describe the relevant part of the sensed values, whereas LSCs translate its less significant parts.

We are now able to present the authentication scheme in WSN based on chaotic iterations, as it has been introduced in [18, 21, 22]. Let:

- $(K, N) \in [0, 1] \times \mathbb{N}$ be an authentication key,

- $X \in \mathbb{B}^N$ be the N LSCs of a snapshot C of a WSN,

- $(S^n)_{n \in \mathbb{N}} \in [\![1, N]\!]^{\mathbb{N}}$ be a strategy, which depends on the authentication signal $M \in [0, 1]$ and K,

- $f_0 : \mathbb{B}^N \to \mathbb{B}^N$ be the vectorial logical negation.

So the watermarked media is C whose LSCs are replaced by $Y_K = X^N$, where:

$$\begin{cases} X^0 = X \\ \forall n < N, X^{n+1} = G_{f_0}(X^n). \end{cases} \tag{8.4}$$

To sum up, chaotic iterations are realized on the least significant part of the sensed values. $(S^n)_{n \in \mathbb{N}}$ can be generated using a "Chaotic Iterations with Independent Strategy (CIIS)" approach: the strategy is independent from the sensed values C, as follows [18, 21, 22]. Consider another time the Piecewise Linear Chaotic Map (see [126]) defined by:

$$F(x, p) = \begin{cases} x/p & \text{if} & x \in [0; p], \\ (x - p)/(\frac{1}{2} - p) & \text{if} & x \in [p; \frac{1}{2}], \\ F(1 - x, p) & \text{else,} \end{cases} \tag{8.5}$$

where $p \in \,]0; \frac{1}{2}[$ is a control parameter. Then, the general term of the strategy $(S^n)_n$ in CIIS setup is defined by the following expression: $S^n = \lfloor N \times K^n \rfloor + 1$, where:

$$\begin{cases} p \in [0; \frac{1}{2}] \\ K^0 = M \otimes K \\ K^{n+1} = F(K^n, p), \forall n \leq N_0 \end{cases} \tag{8.6}$$

in which $\otimes$ denotes the bitwise exclusive or (XOR) between two floating part numbers (*i.e.*, between their binary digits representation), K is part of the authentication key, and M is the sequence of MSCs.

To prove the efficiency and the robustness of the proposed algorithm, some attacks are applied to the same WSN as in the previous section. For each attack, a similarity percentage with the watermark is computed; this percentage

TABLE 8.8: Zeroing attacks.

UNAUTHENTICATION		AUTHENTICATION	
Size (pixels)	Similarity	Size (pixels)	Similarity
10	99.08%	10	89.81%
50	97.31%	50	54.54%
100	92.43%	100	52.24%

TABLE 8.9: Rotation attacks.

UNAUTHENTICATION		AUTHENTICATION	
Angle	Similarity	Angle	Similarity
5	94.67%	5	59.47%
10	91.30%	10	54.51%
25	80.85%	25	50.21%

is the number of equal bits between the original and the extracted watermark. These results have been formerly obtained in [18].

Zeroing Attack In this kind of attack, some nodes of the WSN are put to 0. In this case, the results in Table 8.8 have been obtained. We can conclude that in case of unauthentication, the watermark still remains after a cropping attack: the desired robustness is reached. In case of authentication, even a small change of the carrier sensed values leads to a very different extracted watermark. In this case, any attempt to alter the WSN will be signaled.

Rotation Attack Let r_θ be the rotation of angle θ around the center $(128, 128)$ of the carrier image. So, the transformation $r_{-\theta} \circ r_\theta$ is applied to the watermarked WSN. The good results in Table 8.9 are obtained.

JPEG Compression A JPEG compression is applied to the sensed values, depending on a compression level. Let us notice that this attack leads to a change of the representation domain (from spatial to DCT domain). In this case, the results in Table 8.10 have been found. A good authentication through a compression-based aggregation is obtained. As for the unauthentication case, the watermark still remains after a compression level equal to 10. This is a good result if we take into account the fact that we use "spatial" embedding.

Gaussian Noise Watermarked images can be also attacked by the addition of a Gaussian noise, depending on a standard deviation. In this case, the results in Table 8.11 have been found.

TABLE 8.10: JPEG compression attacks.

UNAUTHENTICATION		AUTHENTICATION	
Ratio	Similarity	Ratio	Similarity
2	82.95%	2	54.39%
5	65.23%	5	53.46%
10	60.22%	10	50.14%

TABLE 8.11: Gaussian noise attacks.

UNAUTHENTICATION		AUTHENTICATION	
Standard dev.	Similarity	Standard dev.	Similarity
1	74.26%	1	52.05%
2	63.33%	2	50.95%
3	57.44%	3	49.65%

8.2.6 Discussion

In this section, we presented a two-layer secure data aggregation for sensor networks. The first layer is based on data encryption with homomorphic properties that provide the possibility to operate on cypher-text. It prevents the decryption phase at the aggregator layers and saves nodes energy. Existing works have exponential bounds in computation and are not suitable for sensor networks, which we tried to relax in our approach. The proposed scheme permits the generation of shorter encryption asymmetric keys, which is so important in the case of sensor networks. The second layer proposes a watermarking-based authentication scheme. The distinct advantage of this layer is to achieve end-to-end authentication where the sink can directly validate the received data from the sources. The experimental results show that our method significantly reduces computation and communication overhead compared to other works, and can be practically implemented in on-the-shelf sensor platforms.

Part V

Conclusion

Chapter 9

Conclusion

9.1 Synthesis

We have studied in this book the possibility to conceive computer programs that evolve in an unpredictable manner.

In order to realize a rigorous study, a mathematical model was required to describe how to understand the sentence "unpredictable evolution of a program on a computer." To do so, well-defined discrete chaotic dynamical systems that generalize serial and parallel iterations have been studied on Cartesian products of Boolean sets. Such systems model accurately the action of a program on a computer. These iterative systems have first been studied using the discrete mathematics framework, in which unpredictability refers to the nonconvergence of the system.

Thanks to tools and results taken from discrete mathematics, it is possible to deduce two fundamental criteria for making such iterative systems nonconvergent [75]. On the one hand, the iteration function must not be a contraction. On the other hand, the iteration strategy must not be a pseudo-periodic one, meaning that no integer must be forgotten by the chaotic strategy (see [75] for further details). As the second constraint concerns more specifically the provided data (that is, the initial conditions of the iterated system), we have focused the study in this book on contraction functions, because this condition is more accurately related to the program under consideration. A canonical example of nonconvergent iterations has been deduced, namely the chaotic iterations that use the vectorial negation, due to its rich potential in our search for unpredictability.

We thus have given a foundation to the notion of unpredictability, because the sole nonconvergence is not sufficiently rich for the objectives we intended to reach. The mathematical theory of chaos naturally appears as the best candidate, at least to initiate the study of unpredictability of such systems. This is why we reviewed the different notions of mathematical chaos.

The next step consisted in making sure that we can study the chaotic iterations inside the mathematical theory of chaos. The definition of an adequate topology τ was required with a view to the targeted applications in information security. Then a modeling stage was necessary to allow the study of the space on which the chaotic iterations were defined. This latter is infinite uncountable, compact, and complete. Continuity and surjectivity of chaotic iterations have then been established, and their periodic points have then been regarded.

Once this topological framework has been established, it has been possible to investigate the chaos of CIs. The formulation of Devaney has been first established for the vectorial negation, while the sensitivity to the initial conditions has been established without the Banks theorem. The constant of sensibility has been computed too. We thus have investigated whether other iteration functions are able to make the CIs chaotic according to Devaney. A characterization of these latter, elements of the set $\mathcal{C}$, has then been established, which is based on the strong connectivity of an associated graph. We have determined the size of $\mathcal{C}$ and shown that the periodic points of CIs with iteration function $f \in \mathcal{C}$ are countable.

After this study of chaos according to Devaney, we have investigated other forms of disorder: expansiveness, topological mixing, chaos according to Knudsen, Li-Yorke, and so on. On a quantitative level, the constant of expansiveness and the topological entropy have been measured. As the iteration function has to be differentiable to compute the Lyapunov exponent, we have established that CIs with the vectorial negation can be rewritten as the iterations of a piecewise linear function on an interval of the real line $\mathbb{R}$. This rewriting of CIs as an iteration on $\mathbb{R}$ makes it possible to compute the Lyapunov exponent. Furthermore, it allows us to compare these chaotic iterations to other functions used in our targeted applications.

Once the study of the unpredictability of the chaotic iterations was realized, the exploitation of these results was regarded. The major problems that can possibly prevent realizing completely the chaos on machines can be summarized as follows: the mathematical theory of chaos is unknown or unused in computer science, chaos being fundamentally a single ingredient that occurs at a time in a larger program (which has no reason to inherit such a property), the computer is a finite state machine (and thus always finishes to enter into a loop), and this latter does not possess any real number, whereas the most famous chaotic sequences are defined on $\mathbb{R}$. We have resolved all these problems by considering the chaotic iterations: the idea was to consider that CIs is a program implemented on a machine, and that the chaotic strategy can be constructed as the iterations pass, using external data. By doing so, the machine is not a finite states one and only integers are manipulated.

The first application was about information hiding. We have shown that it is possible to conceive a hiding algorithm that is chaotic, in the most rigorous sense, and to proceed such that this chaos is preserved during computations. This chaos-based algorithm can be used both in spatial and in frequency domains. The interest of this first application is majorly a theoretical one: this is a proof of concept, showing the feasibility of our approach and its interest while we research a secure and robust method for information hiding.

The second contribution to the information hiding was to show that the mathematical theory of chaos can be interesting in the security field. We have demonstrated that any algorithm can be rewritten as a discrete dynamical system, and thus it can be evaluated using the mathematical theory of chaos. By doing so, new classes of attacks can be studied. Furthermore, it is possible to prove that a given stego-secure information hiding method is chaotic too, thus reinforcing the confidence that can be put on this latter. To show the effectiveness of this complementary approach for information hiding security, the well known spread spectrum technique has been formalized and studied using the mathematical theory of chaos. We have established that this latter is: chaotic according to Devaney, topologically mixing, and strongly transitive. However, this data hiding scheme is not expansive, which is a lack of security when considering attacks in the CMA framework. To solve this problem, a new information hiding algorithm both stego-secure and chaotic has been presented.

Third, a family of new chaotic pseudorandom bit generators called CI PRNGs has been detailed and evaluated. These generators are based on discrete chaotic iterations. The design goal of these generators was to take advantage of the random-like properties of real valued chaotic maps and, at the same time, secure optimal cryptographic properties. More precisely, the design was initiated by constructing a chaotic system on the integers domain instead of the real numbers domain. The randomness and disorder generated by these algorithms have been widely evaluated. Among other things, probability balance, sensitivity to the initial conditions, occurrence of repetitive patterns, linear complexity, uniform distribution, low correlation, and equal frequency contribution have been investigated. They all lead to the conclusion that these generators can be considered as candidates for a large variety of applications in computer science, and certainly in the security field too.

The next application was about hash functions. The security in this case has been guaranteed by the unpredictability of the behavior of the proposed algorithms. The algorithms derived from our approach satisfy important properties of topological chaos such as sensitivity to initial conditions, uniform distribution (as a result of the transitivity), unpredictability, and expansiveness. The choices made in this first hash function are simple: compression function inspired by SHA-1, negation function for the iteration function, etc. The aim was not to find the best hash algorithm, but to give simple illustrated examples to prove the feasibility of using the new kind of chaotic algorithms in computer science.

Finally, two layers of secure data aggregation for sensor networks have been presented. The first one is based on data encryption with homomorphic properties that provide the possibility to operate on cypher-text. It prevents the decryption phase at the aggregators' layers and saves nodes energy. The proposed scheme permits the generation of shorter encryption asymmetric keys, which is important in the case of sensor networks. The second layer proposes a watermarking-based authentication scheme. The distinct advantage of this layer is to achieve end-to-end authentication where the sink can directly validate the received data from the sources. The experimental results show that our method significantly reduces computation and communication overhead compared to other works, and can be practically implemented in on-the-shelf sensor platforms.

9.2   Perspectives

The following perspectives can be imagined concerning the emerging discipline of chaotic machines.

Theoretically speaking, the impact of the choice of topology must be measured precisely. The discovery of properties exhibited by the chaotic iterations must be continued, among other things by enlarging the study of the various notions of chaos common to each function of $\mathcal{C}$; by doing so, it is perhaps possible to find iteration functions that have qualitative or quantitative properties better than the vectorial negation. Generalized CIs that update a subset of components or that encompass delays should be more systematically considered too, and a complete study of iterative systems based on a discrete mathematics approach could be regarded.

In the framework of wireless sensor networks, the information hiding approach for secure data aggregation can be deepened. Large scale simulations and real-life experiments should be conducted to reinforce the confidence put in this method. Furthermore, the theory of iterative systems can be more largely applied to the wireless sensor networks, among other things in video surveillance and intrusion detection [100, 101].

The possibility to conceive chaotic programs could be extended to new applications. We have shown that a certain number of authors use neural networks to conceive information hiding algorithms or hash functions. Most of the time, these neural networks are coupled with chaotic sequences, and the resulting algorithm is supposed to be chaotic. The authors' intention is to conceive neural networks with chaotic behaviors. In the same spirit, it could be possible to apply this work to define a chaotic symmetric cryptographic scheme [47] (as any block cypher is an iterative system) or to study the predictability of current symmetric cryptosystems. Similarly, it should be possible to conceive polymorph viruses [65] having a chaotic behavior: the state of the system can be the code of the virus, the current term of the strategy will be dependent on the id and the content of the host machine, and the iteration function will define the polymorphism, the modification of the host content, and the determination of the next machine to infect.

Bibliography

[1] Mithun Acharya, Joao Girao, and Dirk Westhoff. Secure comparison of encrypted data in wireless sensor networks. In *WIOPT '05: Proceedings of the Third International Symposium on Modeling and Optimization in Mobile, Ad Hoc, and Wireless Networks*, pages 47–53, Washington, DC, USA, 2005. IEEE Computer Society.

[2] André Adelsbach, Stefan Katzenbeisser, and Ahmad-Reza Sadeghi. A computational model for watermark robustness. In Camenisch et al. [46], pages 145–160.

[3] R. L. Adler, A. G. Konheim, and M. H. McAndrew. Topological entropy. *Trans. Amer. Math. Soc.*, 114:309–319, 1965.

[4] J. Ai and A. A. Abouzeid. Coverage by directional sensors in randomly deployed wireless sensor networks. *Journal of Combinatorial Optimization*, 11(1):21–41, 2006.

[5] I. F. Akyildiz, W. Su, Y. Sankarasubramaniam, and E. Cayirci. Wireless sensor networks: a survey. *IEEE Communications Magazine*, 40(8):102–114, August 2002.

[6] Hani Alzaid, Ernest Foo, and Juan Gonzalez Nieto. Secure data aggregation in wireless sensor network: a survey. In *Proceedings of the sixth Australasian conference on Information security - Volume 81*, AISC '08, pages 93–105, Darlinghurst, Australia, Australia, 2008. Australian Computer Society, Inc.

[7] David Arroyo, Gonzalo Alvarez, and Veronica Fernandez. On the inadequacy of the logistic map for cryptographic applications. *X Reunin Espaola sobre Criptologa y Seguridad de la Informacin (X RECSI)*, 1:77–82, 2008.

[8] Jacques Bahi, Jean-François Couchot, and Christophe Guyeux. Steganography: a class of secure and robust algorithms. *The Computer Journal*, 55(6):653–666, 2012.

[9] Jacques Bahi, Jean-François Couchot, Christophe Guyeux, and Adrien Richard. On the link between strongly connected iteration graphs and chaotic boolean discrete-time dynamical systems. In *FCT'11, 18th Int.*

Symp. on Fundamentals of Computation Theory, volume 6914 of *LNCS*, pages 126–137, Oslo, Norway, August 2011.

[10] Jacques Bahi, Jean-François Couchot, Christophe Guyeux, and Michel Salomon.   Neural networks and chaos: Construction, evaluation of chaotic networks, and prediction of chaos with multilayer feedforward network. *Chaos, An Interdisciplinary Journal of Nonlinear Science*, 22(1):013122–1 – 013122–9, March 2012. 9 pages.

[11] Jacques Bahi, Jean-François Couchot, Christophe Guyeux, and Qianxue Wang. Class of trustworthy pseudo random number generators. In *INTERNET 2011, the 3-rd Int. Conf. on Evolving Internet*, pages 72–77, Luxembourg, Luxembourg, June 2011.

[12] Jacques Bahi, Xiaole Fang, and Christophe Guyeux. An optimization technique on pseudorandom generators based on chaotic iterations. In *INTERNET'2012, 4-th Int. Conf. on Evolving Internet*, pages 31–36, Venice, Italy, June 2012.

[13] Jacques Bahi, Xiaole Fang, and Christophe Guyeux. State-of-the-art in chaotic iterations based pseudorandom numbers generators application in information hiding. In *IHTIAP'2012, 1-st Workshop on Information Hiding Techniques for Internet Anonymity and Privacy*, pages 90–95, Venice, Italy, June 2012.

[14] Jacques Bahi, Xiaole Fang, Christophe Guyeux, and Qianxue Wang. Evaluating quality of chaotic pseudo-random generators. application to information hiding. *IJAS, International Journal On Advances in Security*, 4(1-2):118–130, 2011.

[15] Jacques Bahi, Xiaole Fang, Christophe Guyeux, and Qianxue Wang. On the design of a family of CI pseudo-random number generators. In *WICOM'11, 7th Int. IEEE Conf. on Wireless Communications, Networking and Mobile Computing*, pages 1–4, Wuhan, China, September 2011.

[16] Jacques Bahi and Christophe Guyeux. Hash functions using chaotic iterations. *Journal of Algorithms & Computational Technology*, 4(2):167–181, 2010.

[17] Jacques Bahi and Christophe Guyeux. A new chaos-based watermarking algorithm. In *SECRYPT'10, Int. conf. on security and cryptography*, pages 455–458, Athens, Greece, July 2010. SciTePress.

[18] Jacques Bahi and Christophe Guyeux. A new chaos-based watermarking algorithm. In *SECRYPT'10, Int. conf. on security and cryptography*, pages 455–458, Athens, Greece, July 2010. SciTePress.

[19] Jacques Bahi and Christophe Guyeux. Topological chaos and chaotic iterations, application to hash functions. In *IJCNN'10, Int. Joint Conf. on Neural Networks, joint to WCCI'10, IEEE World Congress on Computational Intelligence*, pages 1–7, Barcelona, Spain, July 2010. Best paper award.

[20] Jacques Bahi, Christophe Guyeux, and Abdallah Makhoul. Efficient and robust secure aggregation of encrypted data in sensor networks. In *SENSORCOMM'10, 4-th Int. Conf. on Sensor Technologies and Applications*, pages 472–477, Venice-Mestre, Italy, July 2010.

[21] Jacques Bahi, Christophe Guyeux, and Abdallah Makhoul. Secure data aggregation in wireless sensor networks. homomorphism versus watermarking approach. In *ADHOCNETS 2010, 2nd Int. Conf. on Ad Hoc Networks*, volume 49 of *Lecture Notes in ICST*, pages 344–358, Victoria, Canada, August 2010.

[22] Jacques Bahi, Christophe Guyeux, and Abdallah Makhoul. Two security layers for hierarchical data aggregation in sensor networks. *IJAACS, International Journal of Autonomous and Adaptive Communications Systems*, *(*):***–***, 2011. Accepted manuscript. To appear.

[23] Jacques Bahi, Christophe Guyeux, Abdallah Makhoul, and Congduc Pham. Secure scheduling of wireless video sensor nodes for surveillance applications. In *ADHOCNETS 11, 3rd Int. ICST Conference on Ad Hoc Networks*, volume 89 of *LNICST*, pages 1–15, Paris, France, September 2011. Springer.

[24] Jacques Bahi, Christophe Guyeux, Abdallah Makhoul, and Congduc Pham. Low cost monitoring and intruders detection using wireless video sensor networks. *International Journal of Distributed Sensor Networks*, 2012, 2012. 11 pages.

[25] Jacques Bahi, Christophe Guyeux, and Michel Salomon. Building a chaotic proven neural network. In *ICCANS 2011, IEEE Int. Conf. on Computer Applications and Network Security*, pages ***–***, Maldives, Maldives, May 2011.

[26] Jacques Bahi, Christophe Guyeux, and Qianxue Wang. A novel pseudo-random generator based on discrete chaotic iterations. In *INTERNET'09, 1-st Int. Conf. on Evolving Internet*, pages 71–76, Cannes, France, August 2009.

[27] Jacques Bahi, Christophe Guyeux, and Qianxue Wang. Improving random number generators by chaotic iterations. Application in data hiding. In *ICCASM 2010, Int. Conf. on Computer Application and System Modeling*, pages V13-643–V13-647, Taiyuan, China, October 2010.

[28] Jacques Bahi, Christophe Guyeux, and Qianxue Wang. A pseudo random numbers generator based on chaotic iterations. application to watermarking. In *WISM 2010, Int. Conf. on Web Information Systems and Mining*, volume 6318 of *LNCS*, pages 202–211, Sanya, China, October 2010.

[29] Jacques Bahi, Abdallah Makhoul, and Christophe Guyeux. Efficient and robust secure aggregation of encrypted data in sensor networks for critical applications. In *RESSACS, Journée thématique PHC/ResCom sur RESeaux de capteurS et Applications Critiques de Surveillance*, Bayonne, France, June 2010. Communication orale.

[30] Jacques M. Bahi and C. Guyeux. A chaos-based approach for information hiding security. *ArXiv e-prints*, May 2010.

[31] Jacques M. Bahi and Christophe Guyeux. Hash functions using chaotic iterations. *Journal of Algorithms & Computational Technology*, 4(2):167–181, 2010.

[32] Jacques M. Bahi and Christophe Guyeux. Topological chaos and chaotic iterations, application to hash functions. In *WCCI'10, IEEE World Congress on Computational Intelligence*, pages 1–7, Barcelona, Spain, July 2010. Best paper award.

[33] Jacques M. Bahi, Christophe Guyeux, and Michel Salomon. Building a chaotic proved neural network. *CoRR*, abs/1101.4351, 2011.

[34] Jacques M. Bahi, Christophe Guyeux, and Qianxue Wang. Improving random number generators by chaotic iterations. Application in data hiding. In *ICCASM 2010, Int. Conf. on Computer Application and System Modeling*, pages V13–643–V13–647, Taiyuan, China, October 2010.

[35] J. Banks, J. Brooks, G. Cairns, and P. Stacey. On Devaney's definition of chaos. *Amer. Math. Monthly*, 99:332–334, 1992.

[36] E. Barker and A. Roginsky. Draft NIST special publication 800-131 recommendation for the transitioning of cryptographic algorithms and key sizes, 2010.

[37] Mauro Barni, Franco Bartolini, and Teddy Furon. A general framework for robust watermarking security. *Signal Processing*, 83(10):2069–2084, 2003. Special issue on Security of Data Hiding Technologies, invited paper.

[38] Sebastiano Battiato, Dario Catalano, Giovanni Gallo, and Rosario Gennaro. Robust watermarking for images based on color manipulation. In Pfitzmann [115], pages 302–317.

[39] Dan Boneh and Matt Franklin. Identity-based encryption from the Weil pairing. *SIAM J. of Computing, extended abstract in CRYPTO'01,* 32(3):586–615, 2003.

[40] Dan Boneh, Eu-Jin Goh, and Kobbi Nissim. Evaluating 2-dnf formulas on ciphertexts. pages 325–341. 2005.

[41] R. Bowen. Entropy for group endomorphisms and homogeneous spaces. *Trans. Amer. Math. Soc.,* 153:401–414, 1971.

[42] R. Bowen. Periodic points and measures for axiom a diffeomorphisms. *Trans. Amer. Math. Soc.,* 154:377–397, 1971.

[43] Christian Cachin. An information-theoretic model for steganography. In *Information Hiding,* volume 1525 of *Lecture Notes in Computer Science,* pages 306–318. Springer Berlin / Heidelberg, 1998.

[44] Yanli Cai, Wei Lou, Minglu Li, and Xiang-Yang Li. Target-oriented scheduling in directional sensor networks. *26th IEEE International Conference on Computer Communications, INFOCOM 2007,* pages 1550–1558, 2007.

[45] H. Cam, S. Ozdemir, P. Nair, D. Muthuavinashinappan, and H. O. Sanli. ESPDA: Energy-efficient secure pattern based data aggregation for wireless sensor networks. *Computer Communication Journal (29),* pages 446–455, 2006.

[46] Jan Camenisch, Christian S. Collberg, Neil F. Johnson, and Phil Sallee, editors. *IH 2006: Information Hiding, 8th International Workshop,* volume 4437 of *Lecture Notes in Computer Science,* Alexandria, VA, USA, July 2007. Springer.

[47] Claude Carlet, Pascale Charpin, and Victor Zinoviev. Codes, bent functions and permutations suitable for des-like cryptosystems. *Des. Codes Cryptography,* 15(2):125–156, 1998.

[48] C. Castelluccia, A. Chan, E. Mykletun, and G. Tsudik. Efficient and provably secure aggregation of encrypted data in wireless sensor networks. *ACM Trans. Sen. Netw.,* 5(3):1–36, 2009.

[49] Claude Castelluccia. Efficient aggregation of encrypted data in wireless sensor networks. In *MobiQuitous,* pages 109–117. IEEE Computer Society, 2005.

[50] F. Cayre and P. Bas. Kerckhoffs-based embedding security classes for woa data hiding. *IEEE Transactions on Information Forensics and Security,* 3(1):1–15, 2008.

[51] F. Cayre, C. Fontaine, and T. Furon. Watermarking security: theory and practice. *IEEE Transactions on Signal Processing*, 53(10):3976–3987, 2005.

[52] R. Chandramouli, S. Bapatla, and K.P. Subbalakshmi. Battery power-aware encryption. *ACM transactions on information and system security*, pages 162–180, 2006.

[53] M. X. Cheng, L. Ruan, and W. Wu. Achieving minimum coverage breach under bandwidth constraints in wireless sensor networks. *in IEEE IN-FOCOM*, 2005.

[54] Jung Hee Cheon, Woo-Hwan Kim, and Hyun Soo Nam. Known-plaintext cryptanalysis of the Domingo-Ferrer algebraic privacy homomorphism scheme. *Inf. Process. Lett.*, 97(3):118–123, 2006.

[55] R.C.C. Cheung, N.J. Telle, W. Luk, and P.Y.K. Cheung. Secure encrypted-data aggregation for wireless sensor networks. *IEEE Trans. on Very Large Scale Integration Systems*, 13(9):1048–1059, 2005.

[56] Pedro Comesaña, Luis Pérez-Freire, and Fernando Pérez-González. Fundamentals of data hiding security and their application to spread-spectrum analysis. In *IH'05: Information Hiding Workshop*, pages 146–160. Lectures Notes in Computer Science, Springer-Verlag, 2005.

[57] W. A. Coppel. The solution of equations by iteration. *Mathematical Proceedings of the Cambridge Philosophical Society*, 51(01):41–43, 1955.

[58] Ingemar Cox, Matt L. Miller, and Andrew L. Mckellips. Watermarking as communications with side information. In *Proceedings of the IEEE*, pages 1127–1141, 1999.

[59] Christopher Cramer, Erol Gelenbe, and Hakan Bakircioglu. Video compression with random neural networks. *Neural Networks for Identification, Control, and Robotics, International Workshop*, 0:0476, 1996.

[60] Zhao Dawei, Chen Guanrong, and Liu Wenbo. A chaos-based robust wavelet-domain watermarking algorithm. *Chaos, Solitons and Fractals*, 22:47–54, 2004.

[61] Robert L. Devaney. *An Introduction to Chaotic Dynamical Systems*. Addison-Wesley, Redwood City, CA, 2nd edition, 1989.

[62] Josep Domingo-Ferrer. A provably secure additive and multiplicative privacy homomorphism. In *ISC '02: Proceedings of the 5th International Conference on Information Security*, pages 471–483, London, UK, 2002. Springer-Verlag.

[63] G. S. El-Taweel, H. M. Onsi, M. Samy, and M. G. Darwish. Secure and non-blind watermarking scheme for color images based on dwt. *ICGST International Journal on Graphics, Vision and Image Processing*, 05:1–5, April 2005.

[64] Peng Fei, Qiu Shui-Sheng, and Long Min. A secure digital signature algorithm based on elliptic curve and chaotic mappings. *Circuits Systems Signal Processing*, 24, No. 5:585–597, 2005.

[65] Eric Filiol. *Les virus informatiques : techniques virales et antivirales avances*. 2009.

[66] Enrico Formenti. *Automates cellulaires et chaos : de la vision topologique la vision algorithmique*. PhD thesis, École Normale Suprieure de Lyon, 1998.

[67] Enrico Formenti. *De l'algorithmique du chaos dans les systèmes dynamiques discrets*. PhD thesis, Universit de Provence, 2003.

[68] Nicolas Friot, Christophe Guyeux, and Jacques Bahi. Chaotic iterations for steganography - stego-security and chaos-security. In Javier Lopez and Pierangela Samarati, editors, *SECRYPT'2011, Int. Conf. on Security and Cryptography. SECRYPT is part of ICETE - The International Joint Conference on e-Business and Telecommunications*, pages 218–227, Sevilla, Spain, July 2011. SciTePress.

[69] T. Furon. Security analysis, 2002. European Project IST-1999-10987 CERTIMARK, Deliverable D.5.5.

[70] Teddy Furon. A survey of watermarking security. In Mauro Barni, Ingemar J. Cox, Ton Kalker, and Hyoung Joong Kim, editors, *IWDW*, volume 3710 of *Lecture Notes in Computer Science*, pages 201–215, Siena, Italy, September 15-17 2005. Springer.

[71] J. Girao, M. Schneider, and D. Westhoff. Cda: Concealed data aggregation in wireless sensor networks. In *Proceedings of the ACM Workshop on Wireless Security*, 2004.

[72] Oded Goldreich. *Foundations of Cryptography: Basic Tools*. Cambridge University Press, 2007.

[73] C. Guyeux, N. Friot, and J. M. Bahi. Chaotic iterations versus Spread-spectrum: chaos and stego security. *ArXiv e-prints*, May 2010.

[74] Christophe Guyeux. *Le désordre des itérations chaotiques et leur utilité en sécurité informatique*. PhD thesis, Université de Franche-Comté, 2010.

[75] Christophe Guyeux. *Le désordre des itérations chaotiques - Applications aux réseaux de capteurs, à la dissimulation d'information, et aux fonctions de hachage.* Éditions Universitaires Européennes, 2012. ISBN 978-3-8417-9417-8. 362 pages. Publication de la thse de doctorat.

[76] Christophe Guyeux and Jacques Bahi. An improved watermarking algorithm for internet applications. In *INTERNET'2010. The 2nd Int. Conf. on Evolving Internet*, pages 119–124, Valencia, Spain, September 2010.

[77] Christophe Guyeux and Jacques Bahi. A topological study of chaotic iterations. application to hash functions. In *CIPS, Computational Intelligence for Privacy and Security*, volume 394 of *Studies in Computational Intelligence*, pages 51–73. Springer, 2012. Revised and extended journal version of an IJCNN best paper.

[78] Christophe Guyeux, Nicolas Friot, and Jacques Bahi. Chaotic iterations versus spread-spectrum: chaos and stego security. In *IIH-MSP'10, 6-th Int. Conf. on Intelligent Information Hiding and Multimedia Signal Processing*, pages 208–211, Darmstadt, Germany, October 2010.

[79] D. Hankerson, A. Menezes, and S. Vanstone. *Guide to Elliptic Curve Cryptography.* Springer Professional Computing, 2004.

[80] W. Haodong, S. Bo, and L. Qun. Elliptic curve cryptography-based access control in sensor networks. *International Journal of Security and Networks*, 1(3-4):127–137, 2006.

[81] T. He, S. Krishnamurthy, J. A. Stankovic, T. Abdelzaher, L. Luo, R. Stoleru, T. Yan, L. Gu, J. Hui, and B. Krogh. Energy-efficient surveillance system using wireless sensor networks. *in MobiSys*, pages 270 – 283, 2003.

[82] A. Houmansadr, N. Kiyavash, and N. Borisov. Rainbow: A robust and invisible non-blind watermark for network flows. In *NDSS09: 16th Annual Network and Distributed System Security Symposium*, 2009.

[83] Shih-I Huang, Shiuhpyng Shieh, and J. Tygar. Secure encrypted-data aggregation for wireless sensor networks. *Wireless Networks*.

[84] R. J. Jenkins. ISAAC. *Fast Software Encryption*, pages 41–49, 1996.

[85] T. Kalker. Considerations on watermarking security. In *Multimedia Signal Processing, 2001 IEEE Fourth Workshop on*, pages 201–206, 2001.

[86] Ali Kanso and Nejib Smaoui. Logistic chaotic maps for binary numbers generations. *Chaos, Solitons & Fractals*, 40(5):2557–2568, 2009.

[87] Andrew D. Ker. Batch steganography and pooled steganalysis. In Camenisch et al. [46], pages 265–281.

[88] Auguste Kerckhoffs. La cryptographie militaire. *Journal des sciences militaires*, IX:5–83, January 1883.

[89] D. E. Knuth. *The Art of Computer Programming, Volume 2: Seminumerical Algorithms*. Addison-Wesley, 1998.

[90] Pierre L'Ecuyer and Richard J. Simard. TestU01: A C library for empirical testing of random number generators. *ACM Trans. Math. Softw*, 33(4), 2007.

[91] A. K. Lenstra and E. R. Verheul. Selecting cryptographic key sizes. *Jour. of the International Association for Cryptologic Research*, 14(4):255–293, 2001.

[92] T. Y. Li and J. A. Yorke. Period three implies chaos. *Amer. Math. Monthly*, 82(10):985–992, 1975.

[93] Yantao Li, Shaojiang Deng, and Di Xiao. A novel hash algorithm construction based on chaotic neural network. *Neural Computing and Applications*, pages 1–9, 2010.

[94] Hua-Yi Lin and Tzu-Chiang Chiang. Cooperative secure data aggregation in sensor networks using elliptic curve based cryptosystems. In *CDVE'09: Proceedings of the 6th international conference on Cooperative design, visualization, and engineering*, pages 384–387, Berlin, Heidelberg, 2009. Springer-Verlag.

[95] An Liu and Peng Ning. Tinyecc: A configurable library for elliptic curve cryptography in wireless sensor networks. In *7th International Conference on Information Processing in Sensor Networks (IPSN 2008)*, pages 245–256, April 2008.

[96] Hai Liu, Pengjun Wan, and Xiaohua Jia. Maximal lifetime scheduling for sensor surveillance systems with k sensors to one target. *IEEE Transactions on Parallel and Distributed Systems*, 17(12):1526–1536, 2006.

[97] Shao-Hui Liu, Hong-Xun Yao, Wen Gao, and Yong-Liang Liu. An image fragile watermark scheme based on chaotic image pattern and pixel-pairs. *Applied Mathematics and Computation*, 185:869–882, 2007.

[98] Huadong Ma and Yonghe Liu. Some problems of directional sensor networks. *International Journal of Sensor Networks*, 2(1-2):44–52, 2007.

[99] Matthew V. Mahoney. Fast text compression with neural networks. In *Proceedings of the Thirteenth International Florida Artificial Intelligence Research Society Conference*, pages 230–234. AAAI Press, 2000.

[100] Moufida Maimour, CongDuc Pham, and Doan B. Hoang. A congestion control framework for handling video surveillance traffics on wsn. In *CSE (2)*, pages 943–948. IEEE Computer Society, 2009.

[101] Abdallah Makhoul, Rachid Saadi, and CongDuc Pham. Coverage and adaptive scheduling algorithms for criticality management on video wireless sensor networks. In *ICUMT*, pages 1–8. IEEE, 2009.

[102] H. S. Malvar and D. Florencio. Improved spread spectrum: A new modulation technique for robust watermarking. *IEEE Trans. Signal Proceeding*, 53:898–905, 2003.

[103] G. Marsaglia.   Diehard: a battery of tests of randomness. *http://stat.fsu.edu/ geo/diehard.html*, 1996.

[104] G. Marsaglia. Xorshift rngs. *Journal of Statistical Software*, 8(14):1–6, 2003.

[105] A. Menezes, Paul C. van Oorschot, and S. Vanstone. *Handbook of Applied Cryptography*. CRC Press, 1997.

[106] Thomas Mittelholzer. An information-theoretic approach to steganography and watermarking. In Pfitzmann [115], pages 1–16.

[107] Aidan Mooney, John G. Keating, and Ioannis Pitas. A comparative study of chaotic and white noise signals in digital watermarking. *Chaos, Solitons and Fractals*, 35:913–921, 2008.

[108] S. Oh, P. Chen, M. Manzo, and S. Sastry. Instrumenting wireless sensor networks for real-time surveillance. *Proc. of the International Conference on Robotics and Automation*, 2006.

[109] G. Paul and S. Maitra. *Rc4 Stream Cipher and Its Variants*. Discrete Mathematics and Its Applications Series. Taylor and Francis, 2011.

[110] F. Peng, S.-S. Qiu, and M. Long. One way hash function construction based on two-dimensional hyperchaotic mappings. *Acta Phys. Sinici.*, 54:98–104, 2005.

[111] L. Pérez-Freire, F. Pérez-Gonzlez, T. Furon, and P. Comesaa. Security of lattice-based data hiding against the known message attack. *IEEE Trans. on Information Forensics and Security*, 1(4):421–439, Dec 2006.

[112] Luis Perez-Freire, Pedro Comesana, Juan Ramon Troncoso-Pastoriza, and Fernando Perez-Gonzalez. Watermarking security: a survey. In *LNCS Transactions on Data Hiding and Multimedia Security*, 2006.

[113] Luis Perez-Freire, F. Prez-gonzalez, and Pedro Comesaa. Secret dither estimation in lattice-quantization data hiding: A set-membership approach. In Edward J. Delp and Ping W. Wong, editors, *Security, Steganography, and Watermarking of Multimedia Contents*, San Jose, California, USA, January 2006. SPIE.

[114] A. Perri, R. Szewczyk, J. D. Tygar, V. Wen, and D. E. Culler. Spins: security protocols for sensor networks. *Wireless Networking*, 5(2):521–534, 2002.

[115] Andreas Pfitzmann, editor. *IH'99: 3rd International Workshop on Information Hiding*, volume 1768 of *Lecture Notes in Computer Science*, Dresden, Germany, September 29 - October 1. 2000. Springer.

[116] Congduc Pham and Abdallah Makhoul. Performance study of multiple cover-set strategies for mission-critical video surveillance with wireless video sensors. In *6th IEEE Int. Conf. on Wireless and Mobile Computing, Networking and Communications, wimob'10*, pages 208–216, 2010.

[117] Congduc Pham, Abdallah Makhoul, and Rachid Saadi. Risk-based adaptive scheduling in randomly deployed video sensor networks for critical surveillance applications. *Journal of Network and Computer Applications*, 34(2):783–795, 2011.

[118] Bartosz Przydatek, Dawn Song, and Adrian Perrig. Sia: Secure information aggregation in sensor networks. pages 255–265. ACM Press, 2003.

[119] NIST Special Publication 800-22 rev. 1. A statistical test suite for random and pseudorandom number generators for cryptographic applications. August 2008.

[120] David Richeson and Jim Wiseman. Chain recurrence rates and topological entropy. *Topology and its Applications*, 156(2):251 – 261, 2008.

[121] F. Robert. *Discrete Iterations: A Metric Study*, volume 6 of *Springer Series in Computational Mathematics*. 1986.

[122] O. Rudenko and M. Snytkin. Image compression based on the neural network art. *Cybernetics and Systems Analysis*, 44:797–802, 2008.

[123] Sylvie Ruette. *Chaos en dynamique topologique, en particulier sur l'intervalle, mesures d'entropie maximale*. PhD thesis, Universit d'Aix-Marseille II, 2001.

[124] Laurent Schwartz. *Analyse: topologie générale et analyse fonctionnelle*. Hermann, 1980.

[125] Claude E. Shannon. Communication theory of secrecy systems. *Bell Systems Technical Journal*, 28:656–715, 1949.

[126] Li Shujun, Li Qi, Li Wenmin, Mou Xuanqin, and Cai Yuanlong. Statistical properties of digital piecewise linear chaotic maps and their roles in cryptography and pseudo-random coding. *Proceedings of the 8th IMA International Conference on Cryptography and Coding*, 1:205–221, 2001.

[127] Richard Simard and Université De Montréal. Testu01: A software library in ansi c for empirical testing of random number generators., 2002.

[128] Gustavus J. Simmons. The prisoners' problem and the subliminal channel. In *Advances in Cryptology, Proc. CRYPTO'83*, pages 51–67, 1984.

[129] Ian Stewart. *Does God Play Dices?: The Mathematics of Chaos.* Penguin, 1989.

[130] Dan Tao, Huadong Ma, and Liang Liu. Coverage-enhancing algorithm for directional sensor networks. *Lecture Notes in Computer Science*, pages 256–267, November 2006.

[131] M. S. Turan, A Doganaksoy, and S Boztas. On independence and sensitivity of statistical randomness tests. *SETA 2008*, LNCS 5203:18–29, 2008.

[132] S. M. Ulam and J. V. Neumann. On combination of stochastic and deterministic processes. *Amer. Math. Soc.*, 53:1120, 1947.

[133] G. Voyatzis and I. Pitas. Chaotic watermarks for embedding in the spatial digital image domain. *Proceedings of IEEE ICIP*, 2:432–436, 1998.

[134] David Wagner. Cryptanalysis of an algebraic privacy homomorphism. In *Information Security*, volume 2851 of *Lecture Notes in Computer Science*, pages 234–239. Springer Berlin, Heidelberg, 2003.

[135] J. Wang, C. Niu, and R. Shen. Randomized approach for target coverage scheduling in directional sensor network. *Lecture Notes in Computer Science*, pages 379–390, 2007.

[136] Qianxue Wang, Jacques Bahi, Christophe Guyeux, and Xiaole Fang. Randomness quality of CI chaotic generators. application to internet security. In *INTERNET'2010. The 2nd Int. Conf. on Evolving Internet*, pages 125–130, Valencia, Spain, September 2010. IEEE Computer Society Press. Best Paper award.

[137] X. M. Wang, J. S. Zhang, and W. F. Zhang. One-way hash function construction based on the extended chaotic maps switch. *Acta Phys. Sinici.*, 52, No. 11:2737–2742, 2003.

[138] Yong Wang, Kwok-Wo Wong, Xiaofeng Liao, and Tao Xiang. A block cipher with dynamic s-boxes based on tent map. *Communications in Nonlinear Science and Numerical Simulation*, 14(7):3089 – 3099, 2009.

[139] Xianyong Wu, Zhi-Hong Guan, and Zhengping Wu. A chaos based robust spatial domain watermarking algorithm. In *ISNN '07: Proceedings of the 4th international symposium on Neural Networks*, pages 113–119, Berlin, Heidelberg, 2007. Springer-Verlag.

[140] Di Xiao, Xiaofeng Liao, and Yong Wang. Improving the security of a parallel keyed hash function based on chaotic maps. *Physics Letters A*, 373(47):4346 – 4353, 2009.

[141] Jennifer Yick, Biswanath Mukherjee, and Dipak Ghosal. Wireless sensor network survey. *Comput. Netw.*, 52(12):2292–2330, August 2008.

[142] Wei Zhang, Yonghe Liu, Sajal K. Das, and Pradip De. Secure data aggregation in wireless sensor networks: A watermark based authentication supportive approach. *Pervasive and Mobile Computing*, 4(5):658 – 680, 2008.

[143] Y. Zhu and L. M. Ni. Probabilistic approach to provisioning guaranteed qos for distributed event detection. *in IEEE INFOCOM*, 2008.

Index

Milton Keynes UK
Ingram Content Group UK Ltd.
UKHW040101071024
449327UK00019B/724

9 780367 379940